lso by Jean-Marie Chevalier:

o-editor Gallimard)
ES GRANDES BATAILLES DE L' ÉNERGIE

The New Energy Crisis: Climate, Economics and Geopolitics

The New Energy Crisis

Climate, Economics and Geopolitics

Edited by

Jean-Marie Chevalier

Director, Centre de Géopolitique de l'Energie et des Matières Premières, University of Paris-Dauphine

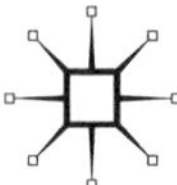

First published 2009 by
PALGRAVE MACMILLAN

Palgrave Macmillan in the UK is an imprint of Macmillan Publishers Limited, registered in England, company number 785998, of Houndmills, Basingstoke, Hampshire RG21 6XS.

Palgrave Macmillan in the US is a division of St Martin's Press LLC, 175 Fifth Avenue, New York, NY 10010.

Palgrave Macmillan is the global academic imprint of the above companies and has companies and representatives throughout the world.

Palgrave® and Macmillan® are registered trademarks in the United States, the United Kingdom, Europe and other countries

ISBN: 978–0–230–57739–8 hardback

This book is printed on paper suitable for recycling and made from fully managed and sustained forest sources. Logging, pulping and manufacturing processes are expected to conform to the environmental regulations of the country of origin.

A catalogue record for this book is available from the British Library.

A catalog record for this book is available from the Library of Congress.

10 9 8 7 6 5 4 3 2 1
18 17 16 15 14 13 12 11 10 09

Printed and bound in Great Britain by
CPI Antony Rowe, Chippenham and Eastbourne

Contents

List of Figures

List of Tables

List of Boxes

Notes on the Contributors

Marie-Claire Aoun is an analyst at the Gas Infrastructure and Network Directorate at the Commission de Régulation de l'Energie (CRE). Previously, she was a fellow reseacher at the Centre de Géopolitique de l'Energie et des Matières Premières (CGEMP) at the Université Paris-Dauphine. She holds a PhD in Economics ('The Oil Rent and the Economic Development of Exporting Countries').

Nadia Campaner is a fellow researcher at the Centre de Géopolitique de l'Energie et des Matières Premières (CGEMP), at the Université Paris-Dauphine. She holds a PhD in Political Science from Paris-Sorbonne Nouvelle ('Energy Interdependences between Russia and the European Union') and an MSc degree from Stockholm University.

Jean-Marie Chevalier is Professor of Economics at Université Paris-Dauphine and Director of Centre de Géopolitique de l'Energie et des Matières Premières (CGEMP). He is also a senior associate with the Cambridge Energy Research Associates (CERA) and a member of the Council of Economic Analysis of the French Prime Minister. He has published a number of books and articles on industrial organisation and energy. His latest book is *Les grandes batailles de l'énergie.*

Michel Cruciani is associate with the Centre de Géopolitique de l'Energie et des Matières Premières (CGEMP). He is also an independent consultant working on climate change issues. He has had long experience with Electricité de France, Gaz de France and the French Trade Union CFDT (Federation of Energy Workers). He was elected member of the board of Gaz de France. His interests are the action of the European institutions and the questions related to environment, energy efficiency and the development of renewables. He graduated from the Ecole Nationale Supérieure d'Arts et Métiers.

Patrice Geoffron is Professor of Economics at the Université Paris-Dauphine and vice-president for International Relations. He is co-director of the Centre de Géopolitique de l'Energie et Matières Premières (CGEMP). His main area of research is the industrial organisation of network industries.

Askar Gubaidullin works as a Project Manager in oil and gas for S2M (France). He holds a PhD in Energy Technology from the Royal Institute of Technology of Sweden and an MSc degree from the Moscow State University.

Iva Hristova is a PhD student at the Centre de Géopolitique de l'Energie et des Matières Premières (CGEMP) at the Université Paris-Dauphine ('The Kyoto Protocol and the Clean Development Mechanism: Impact on Developing Countries').

Jan Horst Keppler is Professor of Economics at the Université Paris-Dauphine and senior researcher at the Centre de Géopolitique de l'Energie et des Matières Premières (CGEMP). He held previous appointments with the International Energy Agency (IEA) as well as the Organisation for Economic Cooperation and Development (OECD). He has published widely on energy and carbon economics.

Delphine Lautier is Professor of Finance at the Université Paris-Dauphine and Associate Research Fellow at Ecole des Mines ParisTech and at (DRM – Cereg) CNRS. Her main areas of research are energy derivative markets and the term structure of commodity prices. She has published a number of books and articles on that topic.

Claude Mandil is former executive director of the International Energy Agency (IEA).

Sophie Méritet is Assistant Professor in Economics at the Université Paris-Dauphine and is a senior fellow of the Centre de Géopolitique de l'Energie et des Matières Premières (CGEMP). She has published several articles on the deregulation process in the electricity and natural gas industries in the US, Europe, Brazil and Mexico. She holds a PhD in Economics at Dauphine University.

Nadia S. Ouédraogo is a PhD student at the Centre de Géopolitique de l'Energie et des Matières Premières (CGEMP) at the Université Paris-Dauphine ('Impact of the Oil Prices on Economic Development in African Countries').

Stéphane Rouhier is a PhD student at the Centre de Géopolitique de l'Energie et des Matières Premières (CGEMP) at the Université

Paris-Dauphine ('Environmental Impact of Rising Energy Use in China: Solutions for a Sustainable Development').

Fabienne Salaün is Associate Professor in Economics at the Université Paris-Dauphine and a senior fellow of the Centre de Géopolitique de l'Energie et des Matières Premières (CGEMP). She holds a PhD in Economics and previously has held positions as expert or manager in Electricité de France before joining EDF Corporate Strategy Division where she is in charge of regulatory issues.

Yves Simon is Professor of Finance at the Université Paris-Dauphine, (DRM – Cereg) CNRS and Editor at Economica, where he is responsible for the management and finance series. He is specialised in international finance and derivative markets. He has published a number of books and articles on that subject.

C. Pierre Zaleski is General Delegate of the Centre de Géopolitique de l'Energie et des Matières Premières (CGEMP) at Université Paris-Dauphine and President of the Polish Historical and Literary Society. He is active in many international scientific institutions such as Moscow International Energy (Vice President), International Academy of Nuclear Energy (past President) and others. He is also member of Polish Academy of Science and Letters and of the European Academy of Arts, Sciences and Humanities. He is author of numerous publications concerning energy issues.

Foreword

The worldwide energy scene is facing a triple crisis. First of all, there is a crisis of supply: the existing and foreseeable capacity cannot assure the supply of energy under all circumstances, taking into account the growth of random phenomena, whether they are technical, accidental, political or meteorological. Secondly, there is the climate crisis: the present tendencies of consumption and production of energy lead inevitably to the emission of greenhouse gases which would amount, in 2050, to five times more than the IPCC – a consensus of the world's scientists – considers acceptable. And finally, there is the economic crisis: the spectacular increase in and the volatility of energy prices contribute to a slowing down of worldwide economic activity and drive the poorest countries into desperate circumstances.

An effort to remain analytic has often led to the treatment of these three crises independently, or even in mutual opposition. It has often been said that one must choose between economic growth and combating climate change, or that the liberalisation of the European electricity and gas markets could only compromise environmental security.

The New Energy Crisis offers an alternative. Beyond the observation that supplies are uncertain, that the emissions of CO_2 are skyrocketing and that prices are increasing, Jean-Marie Chevalier and other contributors study the causes and propose remedies. And here we discover that the causes are similar and, therefore, the remedies are largely the same.

We are not confronted with three distinct crises, but rather one unique energy crisis, created by the thirst for energy, in part, but not only, in the so-called emerging countries, but also by the increase of uncertainties which penalise investment, by state nationalism and by the ineffectiveness of public opinion. If differences appear, they are, for the most part, due to history, geography and geology.

An understanding of the situation should, therefore, be fashioned not by analysing each problem, but rather by analysing each country or homogeneous group of countries, in order to comprehend the differences in approach to a common problem. That is what is done in this work and that is what makes it indispensable for understanding and overcoming the paradox with which the world is today confronted. This paradox derives from an oxymoron: global nationalism.

Never has globalisation been so evident, for the best – growth – or for the worst – the propagation of crises. And also, never has it been so necessary, since energy is now transported over the entire globe and greenhouse gases ignore all frontiers. Globalisation is universally admitted, judging by the blossoming of international forums and conferences devoted to it. Yet, in reaction, one country after another is retreating into a narrow and timid nationalism, building protective walls around their 'national champions', privileging their immediate interests, or their perceptions of them, in the fight against climate change, using energy as a diplomatic weapon, without excluding the possibility that it may become a weapon in the real sense.

Certainly politicians talk about energy a great deal. But do they listen to each other? Do they try to understand the difficulties and the specific challenges that each of them must face up to? One can suspect and fear that they listen only to themselves. One of the saddest examples is that of the energy relationship between the European Union and Russia, which had every reason to be harmonious and mutually beneficial but which has become a source of conflict due to the failure of each party to take into consideration the point of view of the other.

All this needs to be understood. This book and, in particular, the chapters dealing with geographical analysis, imparts this knowledge and provides the necessary enlightenment. It is hoped that it will be read by all those who, the world over, have international responsibilities in the domains of energy and the environment.

Claude Mandil
Former Executive Director,
International Energy Agency (IEA)

Introduction

Jean-Marie Chevalier

This book is about the new energy crisis. The new energy crisis is not related to high oil prices or to the exhaustion of oil and gas reserves. The new crisis comes from the recent intrusion of climate change issues into energy economics and geopolitics. The reality of climate change has been hidden and long denied. Today the warming of the climate is a proven reality and acknowledged by the international scientific community, but no one knows exactly what will be the physical, economic, geopolitical and social impacts of the phenomenon. It could be very costly for the world economy, especially for the more vulnerable countries that are often among the poorest.

Climate change has recently revealed that the current energy/environment equilibrium is unsustainable. The unbalance may be described with a few figures. Today there are 6.5 billion people living on the Earth. Among them 1.2 billion (18 per cent) account for almost 50 per cent of world energy consumption and are responsible for 30 per cent of greenhouse gas emissions (GHG). Among the others, roughly 2 billion people are living on less than two dollars per day. They have access to neither modern energy products (electricity and petroleum products) nor clean water, meaning that they do not have access to economic development. In the West, each US citizen consumes every year eight tons of oil equivalent and he or she does not want to question the American way of life. In the Far East, a Chinese citizen consumes less than one ton of oil equivalent per year but he or she wants more economic growth and more wealth, including car ownership. If the Chinese had today the same standard of living as developed countries they would have 700 million cars, implying an annual gasoline consumption equivalent to the entire annual oil production of the Middle East. This is just impossible. Two other planets would be needed.

On the one hand, millions of people need to increase their energy consumption to feed their economic development. On the other hand, GHG emissions must be reduced to keep the planet clean and acceptable for the coming generations. The situation is aggravated if we take into account demographic factors which will see the world's population rise from 6.5 to 9 billion before 2050, the majority of the newcomers being born in developing countries. The challenge of the century is to provide enough food, water and energy without further damaging the environment: this is what sustainability means.

This inconsistency is what we call 'the equation of Johannesburg' (from the Earth summit of 2002): how is it possible to produce more energy and, at the same time, to reduce emissions significantly? For the first time in human history, we are confronted by the obligation to manage properly a public good, the climate, which belongs collectively to the citizens of the world. But who is going to pay for the proper maintenance of the planet?

The resolution of the equation will come through a combination of three factors: actions, adaptations and higher prices. Actions will be undertaken at various levels:

- At a worldwide level to try to monitor climate change which is a global issue. The Kyoto Protocol was the first attempt at global monitoring. The challenge now is to formulate post-Kyoto regulations. This is the challenge for the 2009 Copenhagen conference.
- At the European level, actions are underway for building a sustainable energy future. One example is the 'Three twenties for 2020', three quantitative targets for 2020 decided in 2007: reducing greenhouse gas emissions by 20 per cent (compared to their 1990 levels), improving energy efficiency by 20 per cent, and increasing to 20 per cent the share of renewable energy in the global energy balance. The European action reflects a great deal of responsibility but it has had little impact on the global current growth of emissions in the world. Based on present trends the IPCC[1] expects CO_2 emissions to grow by a further 45–110 per cent by 2030, with two-thirds of this increase coming from developing countries.
- At national or local levels, energy policies must now integrate climate change. Some countries are already engaged in a real mitigation process, i.e. in a process of emissions reduction. Many others tend to ignore the problem and give priority to economic growth.

Actions will probably be late and insufficient. Late, because the effects of climate change are not yet very visible and it takes time for people to become aware that climate change is an important issue. Late, because the rich are not prepared to change their comfortable daily lives which the poor still dream of. In addition, powerful lobbies are highly efficient at hiding problems and delaying action. Action might also be insufficient because some irreversible changes might already be at work.

Adaptations will be necessary if actions are late and insufficient. The possible effects of climate change are not very well known or evaluated. They concern global and local pollution, sanitary conditions, disease and species extinction, drought, flooding and other climatic catastrophes. Adaptations will be forced by the unexpected effects of climate change that are not evenly distributed. Some populations will have to migrate; some land will disappear or will undergo desertification. For many, especially among the poorest, adaptation will be costly and painful. Some other violent forms of adaptation will happen: famines, epidemics and conflicts and wars for access to land, food, water, energy.

Higher prices for energy goods and for carbon will probably be a variable of adjustment. Significant price increases could take place for different reasons: higher prices and new taxes imposed by energy policies to restrain energy demand, encourage energy efficiency, and reduce emissions and pollution. Higher prices could also be caused by increased scarcity of resources, either because of excessive demand or because of inadequate or delayed investment in high risk countries where oil and gas resources are concentrated. Price evolution will depend, more than in the past, on the geopolitics of the planet for developing and capturing the existing resources. If prices are much higher, the poorest will suffer more and the current income inequalities could be exacerbated. Failure to resolve the equation can be a potential cause of wars, despair and violence.

This book addresses the challenges raised by the new energy crisis. The first chapter sets the stage. World energy consumption is based, for more than 80 per cent, upon oil, coal and natural gas that are, by definition, non-renewable and polluting energy sources. If such a structure prevails, the future becomes unsustainable: unsustainable because global warming is accelerating and because economic growth and the 3 billion newcomers will put pressure on available resources. Access to the resources and their development will exacerbate geopolitical tensions.

For the following chapters we have adopted a regional approach to better understand the dynamics of a multi-power world. Each region has its own specificities in terms of resource endowment, history and sensitivity to climate change. Each region will contribute differently to the history of the century. Each region is both engaged in integrating globalisation, but also sometimes resisting globalisation.

Asia comes first (Chapter 2) because this area represents more than 60 per cent of the world population. The history of this century will fundamentally be determined by what takes place in this area. Asia is at the crossroads of environment/energy issues with growing GHG emissions and a high dependency on oil, gas and coal imports. Special attention is given in this chapter to the three leading countries: China, India and Japan.

The Russian Federation and the newly independent countries of the Caspian Sea region come second (Chapter 3). This scarcely populated, vast area contains huge reserves of oil, natural gas, coal and hydro-resources. Today Russia exports one-third of its gas and two-thirds of its oil to Europe. In the future, Russia and its neighbouring countries may export more energy towards Asia or the United States. This is a place of tensions between conflicting economic and political interests.

Chapter 4 focuses on countries of the South, countries of the lower income categories that are located in South Asia, Africa and Latin America. Most of them are facing energy and economic poverty, even if they have huge local oil resources (the case of Nigeria). Moreover, many of them are very vulnerable to the effects of climate change (such as drought and floods). Population growth is high: the number of people living in Africa (around 1 billion) will double between now and 2050. For these countries the main priority is economic development and the prerequisite is access to energy as a driver of economic development.

Chapter 5 covers the Middle East and North Africa (MENA). Around 66 per cent of world oil reserves and 43 per cent of world gas reserves are located in this area which represents 5 per cent of the world population. Some of these countries are rich or very rich. However, this windfall wealth is unevenly distributed and does not automatically lead to economic development. In fact, many of these countries suffer from the 'resource curse' (more specifically the oil curse). Human development indexes and governance indicators are frequently poor. Climate change is not considered as a real issue and energy prices are heavily subsidised.

Chapter 6 shows that the United States energy policy might be at a turning point. The country accounts for 5 per cent of the world's population and is responsible for roughly 25 per cent of the world energy consumption and related greenhouse gas emissions. Moreover, the country is increasingly dependent on energy imports for oil, oil products and natural gas. However, American citizens are more and more concerned with the inconsistency between their domestic energy model and the global issue of energy/climate change. This country has a great capacity for adaptation and innovation. Will it be in a position to supply part of the answer to the global equation?

Chapter 7, dealing with Europe, is last in position in this geographical *tour d'horizon*. At the beginning of this century, this is a region where twenty-seven countries (plus some close neighbours) are trying, with difficulties, to build a common, responsible and sustainable energy vision for the future. Lower emissions, improved energy efficiency, more renewable energy, and more diversification are the strategic principles that are shaping the future. Europe has introduced the first major market for CO_2 emissions. The impact of European efforts on global warming might be limited but the region has the potential to become a key actor in the resolution of the equation of Johannesburg.

Chapter 8 is devoted to energy finance. Energy covers a wide range of physical products with strong specificities for oil, natural gas, coal and electricity. It also covers a range of financial products, the value of which is more than thirty times higher than the value of physicals. Energy money feeds financial markets. The financial component has contributed strongly to create interdependencies between the various forms of energy, between their physical forms and their financial forms, between the present and the future, and between energy consumption and CO_2 emissions. It has also contributed to create volatility which might be a source of fragility.

In the last chapter we examine how it is possible to overcome the new energy crisis. What can be expected from energy technologies such as nuclear and renewables? Although the world economy has become global, world geopolitics has not followed suit. Nations are still here, defending their wealth, their local interests and their ambitions. Climate is a public resource that needs to be managed in common. But who is going to pay for managing properly the climate? The new energy crisis exacerbates economic and geopolitical tensions. The real challenge of the century is to set up collectively new forms of global regulation to overcome the crisis.

Note

1. The IPCC is the International Panel on Climate Change.

1
The New Energy Crisis

Jean-Marie Chevalier

When current world energy consumption is considered from the perspective of long-term historical trends, it appears that the last 150 years have been an exceptional but unsustainable period: exceptional in terms of the improvement of comfort and standards of living; unsustainable in terms of the climate change which has resulted. Let us take a brief look at the past.

From the dawn of civilisation until the middle of the nineteenth century, man has always used flows of renewable energy: wood, water, wind, human and animal power. For centuries, renewable energies fed a slow but sustainable economic growth. Commercial speed was constant all over the period: the speed of a trotting horse or the speed of a carrier pigeon: about 30 kilometres per hour. World population, which was 430 million in 1500, reached 1 billion around 1820.

From the middle of the nineteenth century until today, the world's population has increased by a factor of six, and GDP by a factor of sixty. The commercial speed is now 1,000 kilometres per hour but it takes only a few seconds to transfer digital information to any place in the world. More than 80 per cent of our energy consumption now comes from fossil, non-renewable and polluting energy sources – coal, oil and natural gas – which have been relatively easily accessible, cheap and abundant. We are now discovering, albeit rather slowly, a disruption in recent evolution: all forms of pollution are severely damaging the planet and the present situation is probably unsustainable and is further aggravated when resource scarcity and demographic growth are taken into account.

Since the first Earth summit in Rio (1992), it has taken more than 15 years for the words 'sustainable' and 'unsustainable' to become more or less accepted by a significant part of the world population, although not by the majority. As a matter of fact, very few people are directly and physically hurt by climate change. Hurricane Katrina in the US, a tsunami in Asia, a heat wave in Europe and a violent monsoon in Asia are local human catastrophes but there is no scientific evidence that they are directly related to climate change. People are reluctant to spend money or to change their daily lives as long as they are not directly affected.

However, the year 2006 appeared to be a turning point in the awareness of the situation. Several elements brought about some sort of crystallisation of the dual energy–environment issue. The International Energy Agency's (IEA) *World Energy Outlook* (2006) begins with the following statement: 'The energy future which we are creating is unsustainable. If we continue as before, the energy supply to meet the needs of the world economy over the next twenty-five years is too vulnerable to failure arising from under-investment, environmental catastrophe or sudden supply interruption.' G8 leaders meeting with the leaders of several major developing countries (China, India, Brazil, South Africa and Mexico, called the 'Plus Five') in St Petersburg endorsed that judgement. Agreeing to act with resolve and urgency, they adopted a Plan of Action and asked the IEA to 'advise on alternative energy scenarios and strategies aimed at a clean, clever and competitive energy future'. At the very same moment the British economist Nicholas Stern published a report (the Stern Review) in which he estimated that the action to now reduce greenhouse gas (GHG) emissions represents a rather modest investment compared to what would be the cost of inaction for the world economy. In France, an official report requested by the government was presented in October 2006. It proposed a target for 2050: dividing by four the level of greenhouse gas emissions as compared to the 1990 level (Boissieu 2006). Even in the United States, which did not ratify the Kyoto Protocol, the question of climate change is now on the agenda. Al Gore's film *An Inconvenient Truth* (2006) is a pedagogical contribution which shows what could be some of the impacts of climate change for certain parts of the planet: it is frightening to imagine what may happen to the Netherlands, Manhattan or Bangladesh if sea levels rise. In 2007, the 4th IPCC[1] Report presented new alarming data on the subject. In 2007, the dual attribution of the Nobel Prize for Peace to Rajendra K. Pachaury, IPCC's chairman and Al Gore is highly symbolic. The international scientific community is calling for urgent action.

What will be the form of economic growth during this century? Are we going to resolve the 'equation of Johannesburg' (more energy, less emissions) to provide more energy for the economic development of the poorest while maintaining a sustainable planet? This first chapter is about the world's economic and energy dynamics. It raises the question of human energy needs and available resources. Then, we will try to analyse, from the present situation, the driving forces that are shaping the future. How strong are the current historical trends and what are their implications and their limits? What are the uncertainties and the risks of the future? Throughout this analysis we will adopt a dialectical approach which emphasises a process of permanent opposition between conflicting positions. The history of this century, with energy and environment as dual key elements, will be shaped through a series of permanent battles and conflicts. These are *Les grandes batailles* of this century (Chevalier 2004).

In a broad sense, opposition comes from economics, politics and culture.

- Economic dynamics divide the rich and the poor (within each nation and worldwide): resource scarcity and prices; short-term profit and long-term benefits; public goods (such as the climate) and private goods; the physical flows (the quiet coal barges on the river as described by Fernand Braudel) vs. the exuberance of financial derivatives.
- Politics and geopolitics address conflicts between nations and conflicts within nations. Conflicts may concern the access to resources (oil, natural gas, uranium, coal, water, land), and the control and the sharing of resources. Domestic conflicts arise from ethnic or religious rivalries and the sharing of public money (oil money for example). National oppositions also reflect the battle between governments and markets for control over the 'commanding heights', to use the expression of Daniel Yergin and Joseph Stanislaw (1998).
- Culture is another field for opposition and perhaps wars: opposition between religions, between ethnic groups, between the culture of globalisation and the determined resistance of some communities.

Beyond these oppositions are people, corporations and institutions.

- People pursue different objectives: searching for a decent standard of living, looking, by any means, for money and power, defending their values and ideas. They are competing to defend or impose their views.
- Corporations compete to increase their market shares and to maximise profits. They compete but sometimes they also enter into collusion to distort competition and gain market power. They also actively lobby to protect their interests. Some of them have deliberately ignored the issue of climate change, thus delaying actions for reducing greenhouse

gas emissions. The reduction of emissions is costly to powerful industries. However, a growing number of corporations are now considering that they bear some social responsibility for the management of the planet.

- Institutions are national governments and parliaments. Nations compete in terms of economic growth and competitiveness, military power and access to natural resources. There are also multinational institutions such as the European Commission, the European Parliament and all the institutions of the United Nations. Institutions reflect the current balance of power. They impose some legal and institutional frameworks for facilitating and, at the same time, limiting the ambitions of people and corporations.

The history of this century, in dealing with the resolution of the 'equation of Johannesburg', will follow a path through these multiple oppositions, tensions and conflicts. Conflict does not necessarily mean that there is a winner. The course this century takes will leave room for negotiations, trade-offs and compromises. For example, the 'battle between governments and markets' or, more generally, the conflict between market mechanisms and institutions, calls for a compromise: market mechanisms favour competition, innovation and value creation but they need to be regulated, to some extent, at various levels. New appropriate forms of regulation are needed and remain to be invented (Chapter 9).

The most recent driving force that will shape our future is the very strong interdependence that has been created between the various regions of the world since the fall of the Berlin Wall. 'The world is flat' as Thomas Friedman (2006) put it, meaning that most places in the world are interconnected through the internet and associated technologies.

1 The world energy balance: an unsustainable evolution

The world energy balance describes the contribution of each energy source to world primary energy demand.[2] At the beginning of the twenty-first century, the main characteristic of the world energy situation is the overwhelming dominance of fossil fuels which contribute more than 80 per cent to the world energy supply: oil 35 per cent, coal 25 per cent, natural gas 21 per cent. Fossil fuels are, by definition, non-renewable and polluting, especially in terms of greenhouse gas emissions. Less than 20 per cent is provided by biomass and waste (10 per cent), nuclear

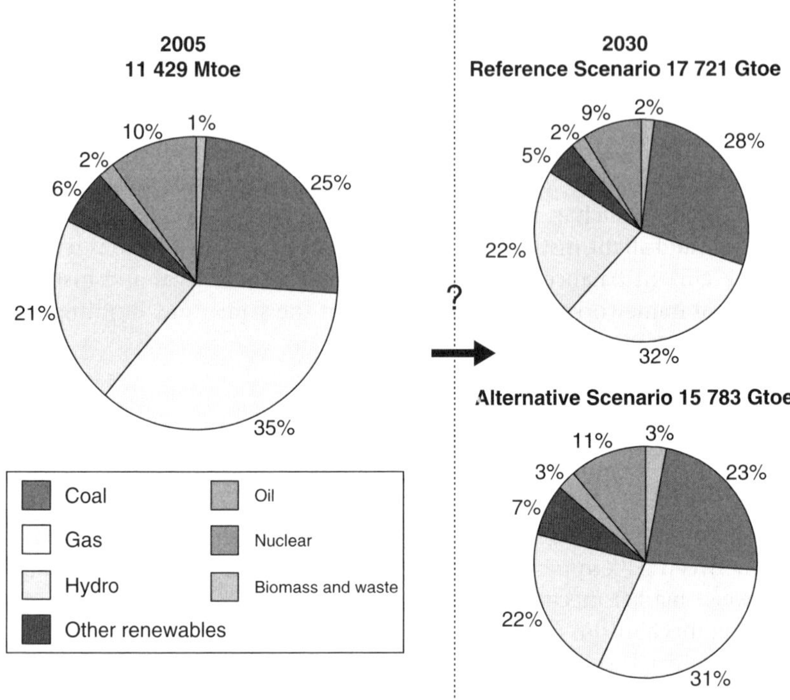

Figure 1.1 The world primary energy balance
Source: CGEMP based on data available from IEA (2007).

(6 per cent), hydro (2 per cent) and other renewables (Figure 1.1). From the current energy balance, the IEA is building scenarios. In the reference scenario, there is no change in energy policies and the projection for 2030 is just unsustainable in terms of GHG emissions and global warming acceleration. An alternative scenario implies some significant policy changes.

1.1 Inertia and rigidities

The structure of the current world energy balance is the historical result of 150 years of rapid growth of fossil fuel consumption to sustain world economic and demographic growth. When the world energy balance is considered on a per capita basis, it shows a great disparity. An average Chinese citizen consumes less than one ton of oil equivalent per year while a US citizen annually consumes about 8 tons of oil equivalent.

The United States, which represents some 5 per cent of the world population, accounts for more than 25 per cent of the global energy consumption and is responsible for 25 per cent of global greenhouse gas emissions. Lying between China and North America, the European citizen consumes about 4 tons of oil equivalent per year. China and other nations from emerging economies need to consume more energy than they do now to feed their economic development, which means more greenhouse gas emissions. We are back to the 'equation of Johannesburg'.

National energy balances differ from country to country. Each country has its own energy structure resulting from domestic resource endowment, national history, level of development and energy policy. In China, the energy balance is dominated by coal (about 70 per cent) with its associated local and global pollution. France represents an exception with the world's highest use of nuclear power (40 per cent). Its Italian neighbour is in a totally different position: there is no nuclear power while oil and natural gas imports account respectively for 42 and 35 per cent of its energy resources. Germany has a much more diversified balance: oil (36 per cent), coal (25 per cent), natural gas (23 per cent) and nuclear (12 per cent).

The world energy balance and national balances reflect a great deal of inertia and rigidity. Associated with energy production and consumption, there are dedicated infrastructures such as oil and gas pipelines, tankers, refineries, gasoline stations, power plants and high voltage and low voltage transmission lines. There are also ships, airplanes, trains and more than one billion cars and trucks. Most of the associated infrastructures have been built over the last 150 years.

Since World War Two, electricity has considerably increased its share in energy systems. In most rich countries, electricity has become an essential good. Any blackout, anywhere in the world, demonstrates the high dependence of modern economies on electricity supply. Even non-electrical heating systems need electricity for ignition, pumping and regulating.

An analysis of energy systems leads us to consider the reasons why human beings consume energy, namely in order to satisfy some very specific needs that vary, in quality and quantity, from one country to another and from one period to another. Five categories of needs can be identified:

- *Need for heat*: low temperature heat (below 100°) for heating, cooking and washing. High temperature heat for production of goods through industrial processes (aluminium, steel, chemicals, etc.).

- *Need for mechanical power*: for the transport of human beings and goods, for industrial processes (to drill, laminate, press, etc.). Mechanical power can be delivered by a number of instruments: steam engines, electrical and internal combustion engines, turbines, and so on.
- *Need for lighting*: progress in lighting – from the old candle to city gas, oil lamps and finally to electric bulbs – has been a decisive factor in accelerating industrialisation, extending working time and improving the domestic standard of living.
- *Need for raw materials*: some primary energy sources are used directly as raw materials such as coke for producing steel, oil and gas that are the basic feedstock for the petrochemical industry (plastics, textiles, fertilisers, synthetic rubber, etc.).
- *Specific needs for electricity* were developed at the end of the nineteenth century when bulbs replaced gas lighting. Then, the rapid development of electrical engines (industry, elevators and household appliances, for example) extended the specific need for electricity. Today high-quality electricity has become an essential good in our daily lives for comfort, work, leisure and transportation.

Starting from human needs, one may analyse the organisation of the various energy value chains and the way they compete. It helps also to identify, for each form of energy, the economic costs and the social costs that are associated in terms of pollution, emissions and other externalities, a question which is crucial for the building of a sustainable future. Let us take three examples: the transport sector, the generation of electricity and city planning.

- *Transportation* has been an important factor over the last 150 years in shaping national and global energy systems and structuring modern economies. Cheap and abundant oil provoked a dash for car and air transport. People were not paying attention to the social cost of transportation systems. Today we are discovering that the transport sector accounts for 14 per cent of greenhouse gas emissions and that the social cost associated with certain modes of transport are much higher than we initially expected. Recent research work emphasises the social costs that are associated with local pollution (air pollution causing diseases and premature deaths, noise and traffic jams) and global pollution. With respect to global pollution, we have to keep in mind that emissions have the same effect wherever they arise from. One ton of CO_2, whether emitted in China or Finland, has the same global effect. Individual choices for mobility have an impact on energy

	Train	Airplane	Car
Total price one-way ticket	**~ 230 Euros**	**Between ~ 80 Euros and 125 Euros**	**~ 60 Euros**
Travel time	2h42	1h15	5h30
CO_2 emissions	12 kg CO_2	52 to 57 kg CO_2	84 kg CO_2

Figure 1.2 Paris–London: cost, time, footprint
Source: Author's calculations based on data available from http://www.voyages-sncf.com/dynamic/_SvMmComparator.

consumption and emissions. The need for mobility can be satisfied, depending on the distance, by walking, bicycling, driving a car or motorcycle, and using public transportation or flying. Transportation illustrates the conflict between energy consumption and GHG emissions. The effects of travelling between Paris and London are indicated in Figure 1.2. Individual choices are determined by price (including taxes), safety and comfort, and also by the value given by each individual to the time element (including the time it takes to travel to the airport or to the railway station).

- *Power generation* has also been an important factor in building modern economies that are more and more dependent on electricity for their daily functioning. Power generation is now responsible for 24 per cent of greenhouse gas emissions. The choice of generating technology is another example which illustrates the economic choices that are related to the competition among various energy value chains. To build a new plant, a company has a choice between various technologies (nuclear, thermal, hydro, solar, wind units) and various energy fuels (coal, natural gas, fuel oil). The final choice is based on the expected cost of generation of a kWh in the new plant, despite the uncertainties surrounding the evolution of costs (capital cost, fuel prices). If social costs are considered, the various competing technologies do not have the same impact on the environment and, therefore, do not have the same cost to society. The evaluation of social costs associated with power generation has been conducted by the European Commission through a long series of studies called 'ExternE' (European Commission 2003). Results indicate that wind, nuclear and, in certain cases, biomass, are the best positioned in terms of low externalities.[3]

- *City planning*. Two cities of the same population, Atlanta in the United States (2.5 million inhabitants) and Barcelona in Spain (2.9 million), illustrate the extreme differences in urban footprint. The first covers an area of 4280 km^2, the second an area of 162 km^2. The annual per capita emission in Atlanta is 7.5 tons of CO_2 while it is 0.7 for Barcelona.[4]

The analysis of energy systems shows that, today, energy and environment are closely related. However, this relation is not yet integrated into individual choices and public policies. The Stern Review states it clearly: 'Those who create greenhouse gas emissions as they generate electricity, power their factories, flare off gases, cut down forests, fly in planes, heat their homes or drive their cars, do not have to pay for the costs of the climate change that results from their contribution to the accumulation of those gases in the atmosphere.' In Stern's terms, this reflects the 'greatest market failure in economic history' (p. 5). We are at the core of the opposition between private goods and one specific public good which is the climate. This is also an invitation to apply the economic principle of internalising externalities, meaning that people have to pay for the damage that they make to climate. However, why should developing countries have to pay to change a current situation for which they bear very little responsibility?

Current energy systems are very inert and rigid. Inertia and rigidity can be observed at two different levels: structure and behaviour. Structure of energy systems covers the organisation of the industry, the current fuel mix and the existence of an energy policy. Infrastructures cannot be transformed rapidly but more easily at the margins. Lead times are long. An interesting but quite exceptional structural change occurred in France, just a few months after the first oil shock (October 1973). In March 1974, the French government decided on a massive programme for the building of nuclear plants in order to decrease French dependence on imported oil. Between 1981 and 2000, 52 nuclear plants were completed with individual capacity from 900 to 1450 MW. In twenty years the structure of the French power system was radically changed. The share of fuel oil, gas and coal in power generation dropped from 50 per cent to 9 per cent while the nuclear contribution increased from 23 to 77 per cent. During the same period French energy 'independence' rose from 27 to 50 per cent.[5]

Behavioural inertia concerns individual and corporate conduct, resistance to change, and the balance of power between private interests and the public authorities which are supposed to defend the long-term public interest. The oil, automobile, aerospace, transportation and power industries would all suffer from any radical change in the patterns of consumption and they frequently, and often efficiently, oppose measures resulting from thc growing concern for climate change.

The performance of energy systems resulting from the building, over 150 years, of structural and behavioural rigidities is very damaging in terms of externalities because of low energy efficiency and huge greenhouse gas emissions. Inertia and rigidities mean that significant changes will be slow and may even be too slow to avoid some irreversibility.

1.2 World energy perspectives: predetermined elements, driving forces, prime movers and uncertainties

Energy experts agree that energy forecasting is nearly impossible because the uncertainties of the future are so numerous that they cannot be totally integrated into a model. In such a context the scenario approach is privileged and widely used by energy companies, governments and international organisations. Scenarios are not forecasts. They result from brainstorming sessions where people of a given organisation try to imagine a small number of 'possible futures'. Each scenario provides an image of a possible future at 20, 30 or 50 years into the future. No probability of occurrence can be given to each of these images but they help to identify basic trends, threats, opportunities and risks. Among the various scenarios of the future that are built, one of them generally reflects a simple extrapolation from the recent past. This 'central' scenario or 'reference scenario' reflects a 'business as usual' (BAU) situation, meaning that no fundamental change occurs in structure and behaviour. The IEA has built a 'reference scenario' and several alternative scenarios. The Cambridge Energy Research Associates (CERA)[6] method for building scenarios is driven by the identification of the various factors that shape the future: predetermined elements, driving forces, prime movers and uncertainties.

Predetermined elements cover some facts and data that are considered to be given at the initial moment. For example, the structure of the current energy systems, with their inertia and rigidities, the structure of the energy industry, the amount of available energy resources (under present knowledge and conditions), and the current state of energy technologies can all be considered as predetermined elements. Other predetermined elements are demographic evolution and the assumptions made about economic growth.

Driving forces are shaping the world energy future. Driving forces are frequently antagonistic, confirming the legitimacy of a dialectical approach. Driving forces may include the economic process of globalisation, the growing concern for the environment and market liberalisation. The new energy crisis introduces an opposition between public goods (the climate) and private interest, between private companies looking for profits and governments or international institutions trying to impose some constraints on public interest.

Prime movers is the term used in scenario building for the actors who are able to alter or to change the rules of the game. They may be companies introducing a major innovation, new governments changing the institutional environment or organisations such as the European Commission or the United Nations.

Uncertainties: the recent evolution of the world energy industry is clearly associated with an increasing complexity: complexity of energy markets with their regulatory frameworks, complexity of the general environment and political and geopolitical complexities. Complexities create uncertainties and risk. Most of the uncertainties are interdependent. The American economist Franc Knight mentioned the 'dynamic uncertainties of the future'. We may refer to these as dialectical uncertainties of the future because they will not only interact between each other, but will also interact with predetermined elements and driving forces. Let us consider four main sources of uncertainties: climate change, economics, institutions and geopolitics.

- *Climate change* is a scientific fact but no one can assert what exactly will be its global and local effects. Warming of the climate system will affect the basic elements of life for people around the world: access to water and water resources, food production, human and animal health, the use of land and its availability, and the environment. Hundreds of millions of people could suffer hunger, water shortages and coastal flooding as the world warms. But the effects of climate change are not evenly distributed. 'All countries will be affected. The most vulnerable – the poorest countries and populations – will suffer earliest and the most, even though they have contributed least to the cause of climate change. The costs of extreme weather, including floods, droughts and storm, are already rising, even in the rich countries.'[7]

  Climate change results from the acceleration of the emissions of greenhouse gases some of which are: carbon dioxide (CO_2), which accounts for 71.4 per cent, methane (CH_4) for 17.5 per cent and nitrous oxide (N_2O) for 10 per cent. Some criticisms were made recently of the current research work on GHG by saying that the role of methane is probably underestimated (Dessus et al. 2008). The sources of emissions are indicated in Figure 1.3: power

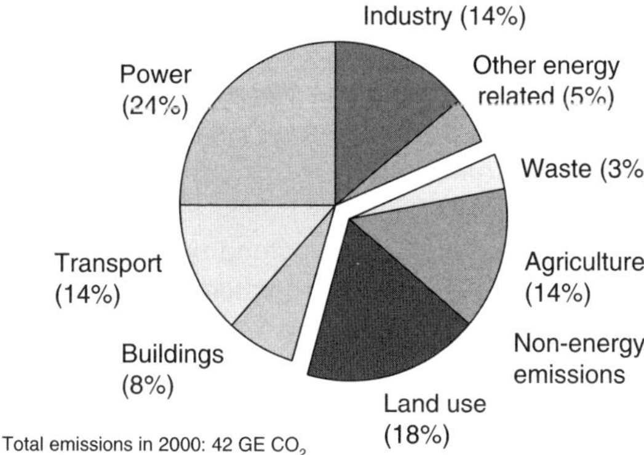

Figure 1.3 Greenhouse gas emissions in 2000 by source
Source: WRI (2006).

generation (especially from coal) comes first, land use (which is basically deforestation) comes second, before the transport sector and agriculture.

A summary of possible severe impacts of climate change is shown in Box 1.1. The impacts of climate change can be anticipated but they cannot be evaluated precisely in quantitative and qualitative terms. Moreover, the time frame and the possible role of amplifying factors are unknown. This means that, step by step, the forthcoming climatic events will put pressure on governments to act more vigorously but the agenda is unknown. The evolution of energy systems is strongly dependent on this dialectical process.

- *Economics*. Globalisation, market liberalisation and the development of new technologies of information and communications have boosted world economic growth. Between 2003 and 2007, world economic growth averaged an annual rate of 4 per cent despite a doubling of oil prices in the same period. Economic growth seems to be insensitive to oil prices. However, the 2008–9 economic crisis may be damaging to the entire interdependent world economy. The world energy demand may peak and even decrease momentarily.
- *Institutional uncertainties*. The liberalisation of the natural gas and electricity markets has given rise to new forms of industrial organisations. Some segments of these industries, considered as natural monopolies,

Box 1.1 Warming will have severe impacts

Climate change threatens the basic elements of life for people around the world – access to water, food production, health, and use of land and the environment.

- Melting glaciers will initially increase flood risk and then strongly reduce water supplies, eventually threatening one-sixth of the world's population, predominantly in the Indian sub-continent, parts of China, and the Andes in South America.
- Declining crop yields, especially in Africa, could leave hundreds of millions without the ability to produce or purchase sufficient food. At mid to high latitudes, crop yields may increase for moderate temperature rises (2–3°C), but then decline with greater amounts of warming. At 4°C and above, global food production is likely to be seriously affected.
- In higher latitudes, cold-related deaths will decrease. But climate change will increase worldwide deaths from malnutrition and heat stress. Vector-borne diseases such as malaria and dengue fever could become more widespread if effective control measures are not in place.
- Rising sea levels will result in an addition of tens to hundreds of millions of people flooded each year with warming of 3 or 4°C. There will be serious risks and increasing pressures for coastal protection in South East Asia (Bangladesh and Vietnam), small islands in the Caribbean and the Pacific, and large coastal cities, such as Tokyo, New York, Cairo and London. According to one estimate, by the middle of the century, 200 million people may become permanently displaced due to rising sea levels, heavier floods, and more intense droughts.
- Ecosystems will be particularly vulnerable to climate change, with about 15–40 per cent of species potentially facing extinction after only 2°C of warming. And ocean acidification, a direct result of rising carbon dioxide levels, will have major effects on marine ecosystems, with possible adverse consequences on fish stocks.

Source: Stern (2006: vi).

are regulated. In many countries independent regulatory authorities have been set up but, frequently, the regulatory systems are not stabilised. Uncertainties concern the changes that could be made to the current forms of regulation. Environmental concerns have also created new forms of regulation in a broader sense: the Kyoto Protocol and all related instruments, the European Emissions Trading System (ETS) with national quota allocations. The evolution of environmental regulation is one of the major uncertainties of our energy future.

- *Geopolitical uncertainties*. A large proportion of oil and gas resources are concentrated in a few countries where political and regulatory stability is at risk. This situation jeopardises the access and the conditions of access to energy resources. It also raises the key question of future investments. Will these countries create the appropriate conditions for access to and development of their energy resources?

The world energy balance is very carbon-intensive. More than 80 per cent of our energy consumption comes from polluting, non-renewable energy resources: oil, coal and natural gas. If nothing is changed, the future is unsustainable because climate change which is already severely damaging the planet will accelerate, with the risk of serious irreversible impacts. There is still time to avoid the worst impacts of climate change if we take strong action now. The Stern Review estimates that if we don't act, the overall costs and risks will be equivalent to losing at least 5 per cent of global GDP each year, now and for ever. In contrast, the cost of action – reducing greenhouse gas emissions to avoid the worst impact of climate change – can be limited to about 1 per cent of global GDP each year. However, not all countries share the same diagnostic, while the urgency and the need for action are global.

2 Geopolitics of energy: wealth, money and power

The geopolitics of energy concerns the balance of power among nations and companies for access to energy resources and, within each nation, the management of energy issues and resources. In oil and gas exporting countries, the geopolitics of energy is closely associated with the appropriation of oil and gas money and its allocation through political decisions. In oil and gas importing countries, the security of energy supply is a major political concern. The geopolitics of energy embraces energy policy, foreign policy and sometimes military action.

Oil and gas have a specific position in the geopolitics of energy for three main reasons: first, because the large consuming areas do not correspond to the large producing areas; second, because oil and gas production and consumption generate huge amounts of money; and third, because the Organisation of Petroleum Exporting Countries (OPEC) has a key role in the determination of the price of oil which is the main benchmark for world energy pricing.

The stake of energy money is illustrated by oil economics with the strange paradox of a very cheap raw material (crude oil) annually generating an incredible amount of wealth. Each year, the world sales of petroleum products, all taxes included, represent about two trillion dollars. The average cost of discovering, producing, transporting, refining and distributing these products can be evaluated at approximately 500 billion dollars. The difference between sales and cost (2000–500) constitutes what we call 'the oil surplus', the amount of wealth which is generated each year by the oil sector, a sum of 1500 billion dollars, the equivalent of the GDP of a country like France. Oil surplus is the 'oil pie' which is shared between oil producing countries, oil consuming countries and all the players that act along the value chain: oil companies, contractors, facilitators, etc. Several factors explain the size of the surplus and the sharing process (see Chapter 5). The cost of oil production all over the world varies from one to fifteen dollars per barrel. That means that low cost producers automatically benefit from a mining rent which is similar to the classical rent analysed by David Ricardo. In addition, OPEC has, at certain moments, an influence on oil price determination, adding a form of monopoly rent over and above mining rents. On the demand side, petroleum products, such as gasoline, diesel oil and jet fuel have no short-term substitutes and therefore enjoy a monopoly situation (which is not the case for fuel oil in competition with natural gas and coal). Governments of developed oil importing countries make use of this monopoly situation to impose taxes. In a country like France, taxes on oil products represent roughly 10 per cent of total state revenues. At the world level, governments of consuming countries take more than 50 per cent of the oil surplus.

Except for oil, energy money is more scattered. The gas surplus is not as large as the oil surplus for at least two reasons: (i) the cost of gas transmission is very high: between seven and ten times more expensive than for oil on a thermal equivalent basis; (ii) natural gas has no monopoly position at the consumer's gate. Gas has to be competitive with its substitute. Nonetheless, gas revenues play an important role in the economies of gas exporting countries that are also suffering from the 'resource curse' (see below).

These various elements shape the energy playing field in a flat world. We will review (i) the economic and energy inequalities; (ii) energy resources, scarcity and prices; (iii) oil and gas exporting countries and resource nationalism; (iv) oil and gas importing countries and security of supply; (v) corporate players; (vi) the world playing field; and (vii) the geopolitics of oil and energy prices.

2.1 Economic and energy inequalities

Energy inequalities have already been mentioned. Per capita annual energy consumption is between 0.5 ton of oil equivalent (toe) in sub-Saharan Africa, and 8 tons in the United States. 1.6 billion people do not have access to modern energy fuels (oil products and electricity), meaning that they do not have access to economic development and they spend a good deal of time collecting local energy resources, such as wood and dung, with all the negative associated effects. In addition, the poorest energy importing countries are directly impacted by high oil prices. The financial burden of the oil bill is high. Very often some oil-fired power plants are partly or totally shut down and some public social programmes have to be cancelled or delayed (see Chapter 4).

Fuel poverty is also a problem in developed countries. In the United Kingdom, for example, fuel-poor households are precisely defined by the British government as those who spend 10 per cent or more of their income on heating. Fuel poverty results from a combination of factors: relatively low income, high fuel price, poor housing conditions characterised by inadequate insulation and inefficient heating systems. About 2 million households are facing fuel poverty in the United Kingdom. The eradication of fuel poverty is an objective of the government. In France, the 'Service public de l'électricité' and the 'Service public du gaz naturel' define precisely how the disadvantaged households must be supplied. From an institutional approach, fuel poverty mainly concerns heating and electricity. However, the question of transport has also to be taken into consideration. In a developed country, a fraction of the low income population, the 'working poor', needs an automobile to go to work. The automobile is frequently a second-hand car with low energy efficiency standards. An increase in the price of gasoline has a significant impact on the purchasing power of the family.

The evolution of energy prices has social impacts on the poor. More generally, higher energy prices, associated with the dynamic of global capitalism, may aggravate the present economic inequalities.

Box 1.2 Millennium Development Goals

In September 2000, 189 countries signed the United Nations Millennium Declaration. In so doing, they agreed on the fundamental dimensions of development, translated into an international blueprint for poverty reduction. This is encapsulated in the Millennium Development Goals that are focused on a target date of 2015:

- Halve extreme poverty and hunger
- Achieve universal primary education
- Empower women and promote equality between women and men
- Reduce under-five mortality by two-thirds
- Reduce maternal mortality by three-quarters
- Reverse the spread of diseases, especially HIV/AIDS and malaria
- Ensure environmental sustainability
- Create a global partnership for development, with targets for aid, trade and debt relief

The United Nations' Development Goals for eradicating poverty are challenging but not easily attainable (Box 1.2). The central messages of the 2008 Global Monitoring Report on the Millennium Development Goals (2008) are clear: 'On current trends, the human development MDGs are unlikely to be met ... Progress toward MDG are slowest in fragile states, even negative on some goals.' The growth of inequalities might be a source of additional tensions and demands.

2.2 Energy resources: scarcity and prices

One of the most important determinants of energy geopolitics is the concentration of oil and gas reserves in a small number of countries that are 'countries at risk'. Access to these reserves is vital for the world economy. More broadly, 80–90 per cent of world reserves of oil and natural gas are in the hands of fewer than thirty oil and gas exporting countries (Figure 1.4).

Many of these countries can be considered as countries 'at risk', meaning that political stability is fragile. The main source of fragility is the 'oil curse' (or more generally the 'resource curse'): oil and gas money distorts economic development and corrupts institutions (see Chapters 4 and 5). Most of these economies, with a high demographic growth rate,

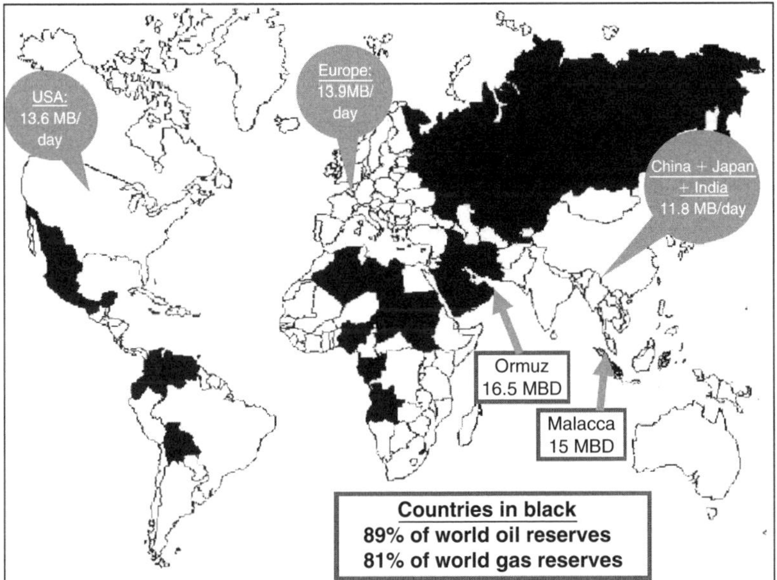

Figure 1.4 Concentration of oil and gas reserves in countries 'at risk'
Source: CGEMP based on data available from BP (2008) and IEA (2008).

are unable to create enough jobs for the young population. Inequalities are growing and if one considers that many of these countries are largely Muslim, one may say that the combination of poverty, frustration and radical Islam may encourage domestic and international terrorism.

Facing the nations who own the reserves are the nations that seek access to the resources (Figure 1.4): Asia has a current low level of energy consumption per capita but a huge potential for economic growth and oil and gas imports; the United States and Europe are more and more dependent on oil and gas imports and cannot rapidly replace non-renewable imported fuels by domestic renewable energy sources. By looking more closely at Figure 1.4, one can see that Asia, with 4 billion inhabitants (60 per cent of the world population), in 2005, and 5.2 billion in 2050, is desperately looking for oil, natural gas and water while, just on the northern frontier, Russia, with its shrinking population of 144 million inhabitants, controls 30 per cent of world gas reserves and 4.5 per cent of world oil reserves and has huge water resources. This map illustrates some of the most important geopolitical stakes of this century.

	Oil	Gas	Coal
Reserves (in Gtoe)	177	160	1907
Ratio reserves/ production (in years) 2007	42	60	133
1973	31	48	na
Concentration of reserves	OPEC = 76%	OPEC = 51% (without Angola and Ecuador) Russia, Iran, Qatar: 25%, 16%, 15%	US = 27% Russia = 19% China = 14% India = 7% Australia = 9%

Figure 1.5 World oil, gas and coal reserves (2007)
Source: CGEMP based on data available from BP *Statistical Review of World Energy*, June 2008.

2.2.1 *Oil reserves*

Every year, data are published on oil and gas reserves by various organisations.[8] They confirm the high concentration of oil reserves in the hands of a few countries (Figure 1.5). If the volume of reserves is divided by the current production, we obtain a ratio (in years) which is frequently misleading and wrongly understood. In 1973, at the first oil shock, the ratio was 31 years and some people were saying that world oil reserves would be exhausted before 2000, taking into account the growth of consumption. Today, the ratio is 42 years. What has happened since 1973? Exploration and production techniques were greatly improved by rapid innovation. The rate of recovery was increased significantly.[9] Huge discoveries were made, especially offshore (in the North Sea, Brazil and along the African coast). A similar evolution can be expected in the future for exploration and production. The concept of proven recoverable reserves has, therefore, a certain elasticity. Reserves can be increased by the combined action of technology, prices and investment. The figures for reserves include some unconventional oil reserves: such as the tar sands of Canada and Venezuela.

However, oil reserves are, by definition, exhaustible and this raises the popular question of 'peak oil'. When will oil production begin to decline? How sharp will the downward slope be? In a time of tight supply, high and volatile oil prices, anxiety about security of supply and environmental concern, the peak oil debate is raging. On the one hand a group which can be called 'the friends of the peak' are sending alarm signals saying that the peak will come very soon for geological reasons and that the world economy is ill prepared for it. On the other hand, a number of experts, among them the Cambridge Energy Research Associates (CERA), present a more detailed analysis, based upon a huge database on oil and gas fields, which indicates that the form and the date of the peak will be determined by a great number of different factors.[10]

According to CERA, a plateau of production will occur – but not in the near future – and supply will not 'run dry' soon thereafter. Global oil production will eventually follow an 'undulating plateau' before declining slowly. Global resources, both conventional and unconventional, are adequate to support production growth and a period on an undulating plateau. CERA holds that above-ground factors (meaning geopolitics) will play the major role in dictating the end of the age of oil. As Daniel Yergin put it 'The main risks are not below the ground but above the ground.'[11] The main question is not the reserves but their development. The appropriate investments to be made could be impeded by political decisions or political turmoil.

When considering the date and the form of peak oil production, another factor is to be taken into consideration: the environmental constraints. With the threats associated with climate change, some large energy consuming countries could be induced to curb oil consumption by improving energy efficiency and by encouraging the substitution of petroleum products by natural gas, biofuels or even electricity (the hybrid refilling car or bus). In that case, peak oil could happen, not because of geological limits or political turmoil but by a peaking of demand for oil products. In the United States, 2007 may well have been the peak for gasoline demand with the beginning of a decline explained by high prices and biofuels substitution (Yergin 2008).

The peak oil problem may be explained by a simple illustration (Figure 1.6). The illustration is simple but it says nothing about the date of the peak because the date and the form of the production curve will, in fact, be modulated by a long series of unpredictable factors:

- The geological limitation of resources (on a field by field basis) in relation to technology improvements.
- The price of oil which will encourage or discourage investment and oil demand.

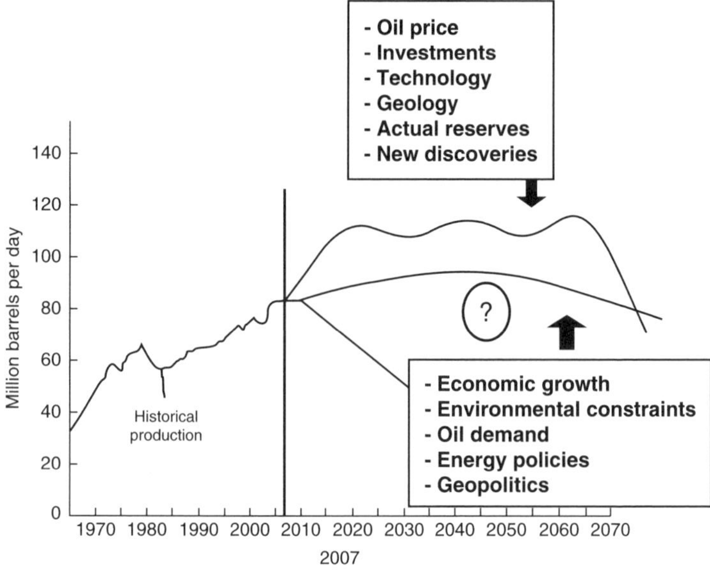

Figure 1.6 What form for peak oil?
Source: CGEMP.

- The policies of oil-rich countries regarding the conditions of access to resources and the development of their productive capacity.
- The energy policies of large oil importing countries: environmental constraints, improvement of energy efficiency and interfuel substitution will all have consequences for the demand for petroleum products.
- The amount and the timing of investments for oil exploration, development and production.
- The impact of technology on exploration, unconventional oil production and field management.
- The rate of recovery which is now about 30–35 per cent could be significantly increased.
- Unpredictable geopolitical and climatic turmoil: wars, civil unrest, accidents, natural catastrophes.

All these factors are qualitatively predictable but their timing and their quantitative evaluation remain totally unknown. The main conclusion is very clear: the end of oil is not for tomorrow but we have to prepare for the post-oil period.

2.2.2 Natural gas reserves

The world reserves of natural gas are slightly less abundant than those of oil. The ratio reserves/production is higher than for oil: 60 years versus 42 years. What was said for oil about this ratio is valid for natural gas, except for the rate of recovery which is already around 80 per cent. However, the world potential for new gas discoveries is probably higher than for oil. This can be explained by the physical and economic differences between oil and gas. Both products have the same chemical components (carbon and hydrogen) but one is liquid, cheap to transport, easily marketable in a global market and has, in addition, a monopoly position on transport fuels. Natural gas transmission is much more expensive. There is no global gas market (the domestic US market is the only competitive integrated market). At the burner tip, gas has to compete with its substitutes and, therefore, final gas demand has to be guaranteed to justify the building of gas lines that are sometimes thousands of kilometres long. This explains why, in the past, oil discoveries were developed almost automatically, while gas discoveries were often ignored and not developed. When natural gas was a by-product of oil production, gas was generally flared on the field. Even today, gas flaring represents a significant share of the world natural gas production and is an important source of carbon emissions.

Although there is an emerging global gas market there is still the question of transmission cost and market risk for the development of gas fields that are far away from consuming areas. In fact there are many gas fields – in Russia, Kazakhstan, Algeria, Iran and Qatar – that contain 'stranded gas' and that will only be developed when demand and prices provide the required incentives.

Ownership of gas reserves is also concentrated in a few hands but the partition is different from that of oil (see Figure 1.5). Russia has the lion's share with about 30 per cent of world gas reserves. Fifty-six per cent of the reserves are found in three countries: Russia, Iran and Qatar. Russia comes first for gas exports (22 per cent) followed by Canada (19 per cent), Norway (11 per cent), Algeria (10 per cent), the Netherlands (7 per cent) and Indonesia (6 per cent).

Russia has a key position in the world geopolitics of gas. For the time being, the majority of Russian gas exports go to the European countries but, in the medium term, if Russian reserves are developed, Russia could export towards Asia and also to the US in the form of LNG (liquefied natural gas). That situation would give Russia an exceptional position of arbitrage between the three largest gas markets. All these elements and the exceptional growth of LNG trading contribute to the development of a global gas market with a growing number of opportunities for swaps, arbitrages and for the development of new gas hubs.

2.2.3 Coal reserves

Coal is the most abundant fossil fuel. On an energy content equivalent, coal reserves are eleven times bigger than the oil reserves (see Figure 1.5). The ratio reserves/production is 133 years. Probable reserves could be much higher. Coal is found in many countries but 80 per cent of reserves are located in six countries: the United States (28 per cent), Russia (18 per cent), China (13 per cent), Australia (9 per cent), India (7 per cent) and South Africa (5 per cent). Coal production is primarily used for local consumption but a few countries export coal. The main exporters of steam coal are Australia, Indonesia, Russia and South Africa. China was a large exporter but became a net importer in 2007.

The situation of uranium resources is described in Box 1.3

Box 1.3 Uranium resources and demand

Known uranium resources (recoverable at up to US$130/kgU) are concentrated in a few countries, but with lower geopolitical risk than oil: Australia, 23 per cent; Kazakhstan, 15 per cent; Russia, 10 per cent; Canada and South Africa, 8 per cent each; USA 6 per cent; and Namibia and Niger, 5 per cent each.

These known resources are estimated at 5.5 million tonnes (Mt). Prognosticated and speculative resources are estimated at 10.5 Mt. Unconventional resources, mainly in phosphate deposits (as by-products), are estimated at 22 Mt. An estimated 4,000 Mt are contained in seawater and granites, but their extraction cost makes them uneconomical for present-generation reactors.

During the past decade, only 50–60 per cent of world uranium consumption was supplied by freshly mined uranium. The remainder, 'secondary resources', came from excess commercial inventories (around 20 per cent), recycling of former military material, and reprocessed spent fuel.

Today, these secondary resources are almost exhausted. Production from existing mines, which was reduced to a minimum, has only a limited capacity to ramp up. Recent price spikes have encouraged a new rush for exploration by hundreds of new entrants, the 'juniors', and some tens of new production projects to be launched in the coming years. We shall soon see whether it is easy or not to add significant uranium supplies to those currently available.

In order to cover reactor requirements, after deduction of available secondary resources, world uranium production must climb from the present level of 40,000 tonnes a year to 60,000 tonnes a year by 2015. There is no resource concern at this point, but increasing world production by 50 per cent is not a minor challenge.

Total conventional resources will last more than 200 years if uranium demand stabilizes at the current level, below 70,000 tonnes a year. However, as many studies have indicated, there may be a need for nuclear power expansion connected with exhaustion of fossil fuels and the negative impact of CO_2 emissions. If one considers the most dynamic development of the 'nuclear renaissance' suggested by some, both conventional and unconventional resources could be exhausted by 2060, assuming that all new reactors built will be of present technology with an improved fuel cycle (1.5 per cent of natural uranium converted to energy) and assuming the constitution of reserves for the operation of existing reactors over their lifetimes (60 to 80 years).

Of course, a colossal effort would be required to ensure this supply. It is therefore clear that in the event that nuclear energy development is very dynamic, the commercial deployment of fast breeder reactors must take place around 2050 or even before. Those reactors would use already mined uranium, left over from the operation of present-technology reactors, and would be able to ensure supply of electricity at a high level (6,000 to 9,000 gigawatts). This type of reactor also could later use thorium or even uranium from seawater and granite, as they convert into energy almost 100 per cent of natural uranium.

C. Pierre Zaleski (CGEMP).

Source: IAEA (2006), INEA (2008), CGEMP (2003), Exane BNP (2008), IEA (2008), OECD (2008).

The world reserves of fossil fuels are huge, even if they are, by definition, exhaustible. In a 'business as usual' scenario (*ceteris paribus*), there is enough oil, natural gas, coal and uranium to feed a robust economic growth. However, there are a number of limiting factors which can create tensions and/or scarcity. (i) Climate change may impose drastic changes in energy policies to reduce the carbon intensity of energy consumption. (ii) A price will have to be paid for carbon emissions and new taxes may be imposed. (iii) The needed investments for developing the existing reserves might be insufficient or delayed. (iv) Political turmoil or resource nationalism may create supply disruptions. In all cases price will be the adjustment variable, with a likely increase of economic and energy inequalities. Let us now analyse the positions of oil and gas exporting countries and oil and gas importing countries.

2.3 Oil and gas exporting countries and resource nationalism

Since the war in Iraq (2003) and the rise of oil prices, resource nationalism has been revived in a number of oil and gas exporting countries. This change is based upon the higher economic value of their oil and gas reserves and also upon a long history of sovereignty over natural resources. One has to keep in mind the history of the oil industry (Chevalier 1975; Yergin 1991). Mexico was the first country, in 1935, after years of imperialist domination, to nationalise its oil industry and to inscribe in its constitution state sovereignty over natural resources. The creation of OPEC in 1960 by Iran, Iraq, Kuwait, Saudi Arabia and Venezuela was the result of long discussions and negotiations for claiming more control over oil production, oil prices and taxation. Most oil exporting countries have created their own state-controlled oil and gas companies.

The most significant oil and gas exporting countries are the 11 OPEC members (the five founding members of 1960 plus the United Arab Emirates, Algeria, Qatar, Libya, Nigeria and Angola) plus Russia, Mexico, Malaysia and Norway. The current and potential role of OPEC will be reviewed later. The main geopolitical issues are (i) resource nationalism, (ii) security of demand, (iii) the oil curse and (iv) potential instability.

2.3.1 Resource nationalism

The resurgence of resource nationalism is based upon the idea that the state has to keep an almost absolute control over oil and gas reserves, including the conditions of access for exploration and production (frequently reserved to national companies), the pace of development and production, the tax regime, domestic pricing measures and the conditions of export (quantities and prices). Several arguments sustain this political will for control. First of all, governments do not want to be accused by the population of selling off national wealth to foreign companies or to national oligarchs. In Russia, after the partial privatisation of the oil sector by Boris Yeltsin, President Putin firmly recaptured state control over the energy industry and made the conditions of entry for foreign companies tougher. Just before the parliamentary and presidential elections (2007–8), it was important to show that the people's wealth was under Kremlin control. It was also politically important to show that 'the Great Russia' was back on the international scene (see Chapter 3).

Another argument for resource nationalism is to put a political limit on exports and to keep reserves for domestic needs and for future generations. This attitude reflects what could be called 'the temptation of creating scarcity'. Why would oil- and gas-rich countries want to develop their resources to feed the inefficient and polluting energy systems of the richest nations? More prosaically, delays in investments will push prices up. The arguments of limited growth, environmental concern and future generations' needs were very well analysed and applied by the Norwegians when they first discovered oil in the North Sea in the early 1970s. The Norwegian parliament report is still one of the most interesting documents concerning the democratic management of domestic resources.[12]

Latin America provides the best illustration of revived nationalism. With higher oil prices, most of the hydrocarbon-rich countries have been taking action to increase the national and state share of oil and gas resources: higher royalties or windfall taxes, a greater role for state-owned energy companies, and tougher conditions for entry and investment for foreign companies. Venezuela, Bolivia and Ecuador are leading this move. In Mexico, the hope for an upstream opening to private international investment has vanished despite a worrying decline in production.

2.3.2 Security of demand

For oil and gas exporting countries, energy security is security of demand. They criticise severely the taxes that some oil importing countries put on petroleum products and the share that they take of the oil surplus. They have long ignored, along with the international oil industry, the question of climate change. Now, they are beginning to worry about security of demand. The growing concern for the environment, all over the world, and the measures that are being taken may accelerate the end of the oil age with the occurrence of a 'peak demand'.

2.3.3 The oil curse

The geopolitics of oil and gas exporting countries is largely determined by the oil curse (Chapters 4 and 5). The structure of trade balances shows that oil and gas sales represent a very high percentage of export revenues. This means that these countries are not able to export anything other than hydrocarbons. Oil and gas sales also represent a large part of government revenues. These economies are heavily dependent on oil prices and oil demand. Money generated by oil and gas activities is shared among all the participants of the industry: the government and its associated groups and agencies, state-owned companies, and national or international private companies. Corruption is frequently associated with the sharing mechanisms. To maintain social peace in the country, part of the money is redistributed, through public expenditures and subsidised tariffs for natural gas, gasoline, diesel oil, electricity and butane. These subsidies, which represent billions of dollars, aggravate the distortions of the economy and encourage energy inefficiency (IEA 2006). When compared to international prices, gasoline and diesel are sold with an 80/90 per cent subsidy in Iran, Algeria and Venezuela. Oil money enables governments to lessen fiscal dependence on the population and to escape, at least partly, democratic legitimacy. Oil money reinforces the power of the ruling class but it might also encourage rebellions and revolts because oil money does not lead to economic development and tends to aggravate inequalities.

2.3.4 Potential instability

Most oil and gas exporting countries are not mature democracies. Oil and gas money and the resource curse are permanent occasions for conflicts. Conflicts may come from inside the country or from outside. Social unrest in Nigeria and a number of civil wars in Africa have been motivated by oil money. The American intervention in Iraq exacerbated rivalries between Shiites, Sunnites and Kurds. The war between Iran and Iraq (1980–8) and the invasion of Kuwait by Iraq illustrate the temptations to violence sustained by massive purchases of weapons. Sophistication of weapons, including chemical and nuclear weapons, is a serious threat to political stability. Another worrying issue is radical Islam which is developing in the Muslim world. The Muslim world[13] is by itself a mosaic of various countries and ethnicities all of which are more or less vulnerable to radical Islam. However, poverty and growing inequalities provide a solid ground for the development of terrorism, both within a given country and abroad. From an energy point of view, risks include the possible emergence of political regimes that limit production, cut exports and stop evolution towards an occidental way of life.

2.4 Oil and gas importing nations and the security of energy supply

The three major oil and gas importing 'blocs' are North America, the European Union and Asia (see Figure 1.4). Besides these blocs, one should not forget a number of poor countries in Africa and Latin America, which are also oil importers facing a financial burden. The history of this century will be clearly influenced by the three dominant blocs that are increasingly oil and gas importers. Their main concerns are (i) energy dependence and (ii) security of energy supply.

2.4.1 *Energy dependence*

Energy dependence is a recent concept. In the nineteenth century, the energy question was basically one of wood availability, and then access to coal mines. Today the functioning of modern economies is based upon continuous energy flows, part of them being imported. Energy dependence has several dimensions. Energy intensity measures the amount of energy that is necessary to produce one unit of gross domestic product (GDP). In the past 30 years, energy intensity, at a worldwide level, was reduced by 35 per cent, reflecting at the same time a substantial improvement in energy efficiency and also a profound change in the structure of production. At the time of the second oil shock (1979–80), two barrels of oil were necessary to produce $1000 of GDP. In 2008, half a barrel is needed to produce the same constant value. Dependence is also fuel-specific. The dependence on electricity reflects an almost continuous need for all the inhabitants of a rich country. The dependence on petroleum products is high for the transport sector and for the army but, as opposed to electricity, petroleum products may be stored. Secure storage is thus a weapon against sudden disruption.

Another dimension of energy dependence is the amount of domestic consumption that is imported and where it is imported from. In this respect, energy dependence is not bad *per se*. If world markets were perfectly competitive, it would be better to buy cheap energy abroad than to consume expensive domestic energy. The development of Japan since the Second World War essentially relied on imported energy. Therefore, 'energy independence' cannot be a target *per se*. However, energy dependence implies some vulnerability. The vulnerability of a given country to oil supply disruption or to oil price shocks can be measured by a series of indicators: such as the oil intensity of the economy and the share of imported oil over total oil use.

The three blocs mentioned above are increasingly dependent on energy imports. Even if these countries encourage energy efficiency and oil and gas substitutes (nuclear, biofuels, wind and solar) they will import more oil and gas and therefore they will compete to have access to oil and gas resources.

2.4.2 *Security of energy supply*

A standard definition of security of supply is a flow of energy supply to meet demand in a manner and at a price level that do not disrupt the course of the economy in an environmentally sustainable manner. The concept is vast, multiform. It has important time, space and social dimensions. It also has some specificities for electricity, oil and natural gas (Chevalier 2006; Keppler 2007; Mandil 2008). What does it mean for the three blocs of the large importing nations?

- In the United States, when George W. Bush was first elected in 2001, energy was already a priority because the oil dependence on foreign supply had been steadily growing since the early 1970s. Bush's answer was simple and supply-sided. In Bush's view, inspired by the oil industry, America needed additional sources of oil and gas supply. On the domestic side, environmental restrictions on exploration and production were to be removed in order to offer new opportunities for the oil industry. Abroad, foreign countries should understand that it is in their interest to open their territories to international investment. He tried to persuade Mexico to do that but failed. Then there was the war in Iraq. Even if oil was never mentioned as a motive for war, the oil question was always in the background. The American position in Saudi Arabia was becoming destablilised and a 'democratic Iraq' would have provided a very good alternative. Iraq is the fourth largest country in the world in terms of oil reserves, and large parts of the country, which are very promising, have not yet been explored and would provide great opportunities for international oil companies. The US invasion has been a complete disaster, however, and will continue to exacerbate the internal tensions in the country and in the region for a long time to come. The miserable Iraqi saga shows that the world is a complex puzzle where ethnic and religious opposition is exacerbated by oil money and resource nationalism. The puzzle calls more for diplomatic action than for new wars. Since the Iraqi failure, the United States has been more concerned with developing domestic non-hydrocarbon resources (biofuel, coal, nuclear) and the strategic priority is to diversify the sources of supply. However, dependence on oil and natural gas imports is still growing rapidly.
- Asia, with the current rate of economic growth in China and India, needs to import more oil, more natural gas and more coal. The required volumes are growing rapidly. Chinese companies are increasingly present in all countries where there are opportunities for developing oil, natural gas and coal. These companies offer better conditions than the major international oil companies; they are not very embarrassed by corruption, civil rights, poor working conditions and issues of environmental responsibility. China is particularly active in

Africa. Angola is now China's first oil supplier. This aggressive strategy could severely toughen competition for access to resources. Another worrying question from Asia is its energy dependence on Pacific Ocean shipping. For China, Japan, Korea and Taiwan, a large proportion of oil, gas and coal imports is unloaded on the Pacific coast. Moreover, 80 per cent of Japan's and South Korea's oil and about half of China's passes through the Straits of Hormuz and Malacca which are vulnerable chokepoints in international shipping. This explains the efforts of some Asian countries to develop new routes of supply to bypass the chokepoints or to reach western sources of oil and natural gas such as Russia, Iran and Kazakhstan.

- Europe is also confronted by its growing dependence on energy imports. If nothing is done for energy efficiency and to improve the competitiveness of domestic supply, about 70 per cent of the Union's energy requirements will be imported in 2030, compared to 50 per cent in 2007. The European Commission has called for a voluntary energy policy aimed at combining competitiveness, security of supply and sustainability. The objectives for 2020 are: cutting greenhouse gas emissions by 20 per cent, improving energy efficiency by 20 per cent and raising the share of renewable energy to 20 per cent. In parallel, the building of a single European market for gas and electricity is expected to reinforce security of supply. On foreign energy policy, Europe is aiming to 'speak with one voice' on international energy and environmental matters and to reinforce cooperation with Europe's main energy suppliers, especially for natural gas: Russia, Algeria and Libya. Dependence on foreign gas supply could be converted to interdependence, mutual interest and cooperation.

Security of energy supply is a national and international issue. Security of transit through chokepoints has to be collectively managed. Hormuz and Malacca were mentioned above. There are also the Suez Canal, the Bab el Mandeb strait, which provides entrance to the Red Sea, and the Bosphorus, which is one major export channel for Russian and Caspian oil. A major tool for security of supply is storage capacity. After the first oil shock and the creation of the International Energy Agency (IEA), strategic stockpiles of oil and oil products were built in member countries (OECD) to offset major disruptions of supply. These disruptions were expected to come from 'risk areas'. In fact, the most important use of these stockpiles was made in the autumn of 2005 after Hurricanes Katrina and Rita shut down 27 per cent of US oil production as well as 21 per cent of US refining capacity. Security of energy supply is, therefore, a question related to geopolitics but also to unexpected climatic events.

2.5 Corporate players

Thousands of private, public, state-owned or state-controlled companies are involved in the energy industry. They are local, national or multinational. Since energy is a very strategic sector, a number of countries in the world have opted for state control of the energy industry. The cases of the United Kingdom and France after the Second World War were symbolic: most of the oil industry (British Petroleum, Total and Elf Aquitaine), coal mining (British Coal and Charbonnages de France), gas transmission and distribution (British Gas and Gaz de France), and power generation and distribution (Central Electricity Generating Board and Electricité de France) were controlled by governments. The United States, with an almost totally private energy industry, was an exception. Following the same philosophy of strategic state control, a number of developing countries and most of the oil and gas exporting countries have nationalised their energy industry and created state-owned energy companies.

In the early 1980s, the battle between governments and markets turned in favour of markets and privatisations. Under the impetus of Margaret Thatcher's government, the UK has almost entirely privatised its energy industry. In continental Europe, the capital structure of many state-controlled companies has been opened up and some European utilities have become multinational companies (Electricité de France, E.On, GDF/Suez). There has been an extensive international process of privatisation, concentration and consolidation of the industry, with, particularly in Europe, the emergence of a gas and power oligopoly where power and gas activities are combined. However, in developing countries, the model of state control remains well entrenched.

From a geopolitical point of view, some corporate players are directly concerned by world geopolitics. These are mainly the international oil companies (IOC), the national oil companies (NOC) and the large equipment suppliers. In the coal business, where the geopolitical stake is lower, the role of several companies still needs to be mentioned.

2.5.1 The international oil companies

The international oil companies represent a small club. A number of mergers have reinforced the concentration: Exxon/Mobil,

Total/Fina/Elf, BP/Amoco/Arco, and Chevron/Texaco. Coming from the 'seven sisters',[14] these companies ruled the oil industry for decades until the first oil shock. Today they are still powerful players, among the largest corporate players in the world. But they face a number of problems, the main one being resource nationalism, with a number of oil-rich countries closing their doors to, or limiting the entry of, international investments. According to the IEA, more than half of the world's proven oil reserves are either closed to foreign companies, or open but under control of the national oil companies. Three countries – Kuwait, Saudi Arabia and Mexico – remain totally closed to upstream oil investment by foreign companies. The shares of large major oil companies (Exxon/Mobil, Shell, BP, Chevron/Texaco, Total, Conoco/Phillips and ENI) in world oil and gas production are respectively 16 and 18 per cent. The dilemma facing the large oil companies, which are making comfortable profits, is that it is difficult to get access to the available reserves where they could spend money. In addition, large corporations are under the scrutiny of financial markets that are looking for short-term profits and low risks and that are also concerned about good business practices. Finally these companies are also facing tough competition from Chinese, Indian, Malaysian and Brazilian competitors.

2.5.2 The national oil companies

National oil companies in oil and gas exporting countries have gained a very significant share of world hydrocarbon production: about 60 per cent for oil and more than 30 per cent for natural gas. This is the core of an important issue: the strategic decisions of these companies for exploration and development and their international ambitions are under the control of their governments, a situation which goes back to resource nationalism. In addition, governments are suspicious vis-à-vis their national companies which control part of the oil and gas money. A national oil company can easily become a state within the state. The case of Venezuela in 2003 is illustrative of the continuous *bras de fer* between national oil companies and their governments. The company PDVSA wanted a large part of the oil money to finance its domestic and international development. President Chavez wanted oil money for his political and social expenditures. The strike at the company was long. It shut down oil production and exports. President Chavez won. About 200,000 strikers were fired. The capability of the industry and its productive capacity were hit very seriously and for a long time. But, resource nationalism won.

2.5.3 Large equipment suppliers

Equipment suppliers compete worldwide for the supply of an extended range of energy equipment from a simple wire to a nuclear plant. Equipment suppliers and other industrial companies are also deeply involved in the Clean Development Mechanism that is a market mechanism of the Kyoto Protocol, providing incentives for investment in clean energy technologies and energy efficiency in developing countries.[15] From a geopolitical point of view, the suppliers and constructors of nuclear plants have a specific position. There are four big companies: Areva/Siemens, proposing the 1,600 MW EPR (*European or Evolutionary Pressurised Reactor*); Westinghouse, now controlled by Toshiba, proposing the AP 1000 and ABWR 1,350; General Electric-Hitachi proposing the ESBWR 1,550 MW; and the Russian company, Rosatom, proposing the VVR 1000.

2.5.4 The large coal companies

Historically the coal industry was organised on a national basis with producers selling their production locally. Since the early 1980s the world coal market has extended very rapidly, especially for steam coal.[16] After a series of mergers and acquisitions, five major producers have established a strong dominant position in the international coal market: Anglo, BHPBilliton, Glencore/Xstrata, Rio Tinto and Drumond. They are all, except Drumond, multi-commodities producers and traders. For production and/or trading, they are responsible for the following shares of the largest coal exporting countries: 67 per cent for Australia, 38 per cent for Indonesia, 40 per cent for Russia, 86 per cent for South Africa, and 82 per cent for Colombia. This high concentration might have an impact on international coal prices.

2.6 The world playing field

Energy is a highly political matter, but energy geopolitics is only one element in a multi-power world which has its own political and economic dynamics. The extension of the market economy, associated with liberalisation and globalisation, has unleashed the expansion of capitalism under many different forms. Generalisation of market economy does not mean 'the end of history' as was expressed by Francis Fukuyama (1992) because it raises new tensions, new oppositions, and new sources of conflict between nations and within nations. Since the Asian crisis, we have entered a new phase of history in which Asia will have a key role.

However, this key role will probably be subordinated to global questions that concern the whole planet: climate change, stability of the world financial system, security of energy supply, resource scarcity (energy, raw materials, food and water), inequalities and ocean shipping security. The simple game of market mechanisms is not able to provide the answers. The world economy needs new forms of multinational regulation. This is a key challenge for this century.

2.6.1 Unleashing capitalism

One of the most powerful driving forces of the world economy is wealth creation, a revival of the old Adam Smith theory. Capitalism is king under many different forms reflecting the diversity of types of capitalism. There are the multinational corporations that publish annual reports, report to their shareholders, establish guidelines for best business practices, pay taxes, actively lobby governments and finance political campaigns. From these companies, financial markets and investors require value creation within a given legal and regulatory framework. But there are millions of other private companies under individual or family control that are pure money makers and are trying to avoid taxation, regulation and other constraints, especially on reporting. Thousands of 'new capitalists' without any *état d'âme* are joining the global playing field of opportunities every year. Some of them are the 'robber barons' of the twenty-first century.[17] There will be wars between capitalisms (Lorenzi 2008). There are also state-controlled companies that represent another form of capitalism: state capitalism (e.g. Gazprom and the Chinese oil companies). In addition, there is the increasingly active financial community looking for any source of profit, especially through financial derivatives, futures markets, arbitrage, hedge funds, sovereign funds and private equity funds. The total of financial assets traded each year has been estimated to be three times the world GDP. The financial community includes thousands of tax havens that are used, under political complicity, to avoid taxes, to allocate profit secretly and to manage money from crime, drugs, corruption and bribery. It is important to bear that in mind when considering energy matters, because energy, as we have seen, is a huge financial stake everywhere in the world and generates an enormous amount of money, part of it being recycled in the international financial system.

The diversity of capitalism is supported by the diversity of states' institutions. Democratic and non-democratic states' institutions are the products of historical evolution which include wars, revolutions, colonialism and struggles for independence. National institutions, at any given moment, reflect a social model which provides an institutional framework for local capitalisms. This institutional model frequently includes some form of wealth reallocation. Capitalism creates wealth which goes primarily to capital owners. State institutions capture part of the wealth for the purpose of reallocation in order to maintain a sustainable social situation. Reallocation takes the form of social security, unemployment allocations, free education, health care, etc. The surge of oil prices between 2004 and 2008 has greatly increased the amount of money captured by oil exporting countries. Part of this money is used to develop sovereign funds which control thousands of billions of dollars. The intervention of sovereign funds brings a new dimension to the structure of global capitalism. The control of some large international industrial or financial corporations is captured by Asian and Arabic sovereign-wealth funds.

States' institutions broadly include the ruling class, the government, the parliament and other state agencies. Their relationship with capitalism is also characterised by diversity. Political power needs money to run the administration, the military and governmental agencies. Money can be raised by public taxation but also through a number of informal channels. Money is the key to gaining power and keeping it. In return for their money, capitalists need state assistance and state support. This is the most common trade-off.

Many state institutions in the world play the role of countervailing power vis-à-vis wild capitalism and its short-term objectives. However, the problem is that, today, many capitalist entities are able to partly or totally escape national control and many questions of social and public interest are becoming global.

2.6.2 Global economy, global questions and local resistance

Globalisation of the economy is a fact which has been reinforced by unleashing capitalism, by the tremendous expansion of the finance industry and by the development of high-speed communications systems. Globalisation is now developed and promoted by companies coming from China, India, Brazil and the Gulf. Asia's contribution to global economic input, which is now about 37 per cent, could reach around 55 per cent in 2030. The centre of gravity of the world economy is changing. However, the global economy shows some fragility. There are several sources of this fragility:

- Globalisation is good for a great number of people but others are suffering. For many people it offers business opportunities, new jobs and wealth. Others are suffering from cut-throat competition, job losses, delocalisation and threats to their social and health protection. This is not a new phenomenon but it is now global, rapid and more brutal.

- Globalisation seems to aggravate inequalities. Wealth is predominantly created by private companies and profits are predominantly distributed to capital owners. Throughout this process, the rich are becoming richer more rapidly than the poor. The case of China is quite significant. Between 1994 and 2004, the Gini coefficient, which measures inequalities, has grown from 40 to 47.
- The growing disconnection between the real world and the financial sphere is a subject of worry. The financial crisis is the first global crisis of the world interconnected economy.
- Climate change, energy and economic development (the equation of Johannesburg) are global questions that have to be resolved globally. If they are not, the world economy is in great danger.
- There is also local resistance to global capitalism: resistance by exclusion from the process of wealth creation and resistance by ideology or religion that does not accept the social and cultural model which is proposed by capitalism. This is probably more complex than the 'clash of civilisations'[18] as viewed by Samuel Huntington. The opposition between the 'Occidental' view and the 'Muslim world' is real but there are other more important oppositions, including opposition within the Muslim world. Greater Asia (including the Middle East and Central Asia) is a mosaic of oppositions between nations, between social models, and between ethnic groups and religions. These multiple oppositions may lead to dialogue and cooperation; they may also trigger the wars of the century.

2.7 Geopolitics of oil and energy prices: the third oil shock

Since oil represents 36 per cent of world energy consumption, oil prices play a leading role in the determination of energy prices. Gas prices roughly follow oil prices. The pricing of coal is more independent. What happened in the world oil markets since 1998 has much to teach us (Figure 1.7).

In 1998, after a long period of price instability, crude oil prices reached their lowest level since 1971 in constant dollars: $10 per barrel. This level was not acceptable to the large oil exporting countries (nor to the high cost marginal producers in the United States). At that price, exporting countries cannot balance their budgets and finance public expenditures. In 1999, OPEC countries met and decided to cut production and to try

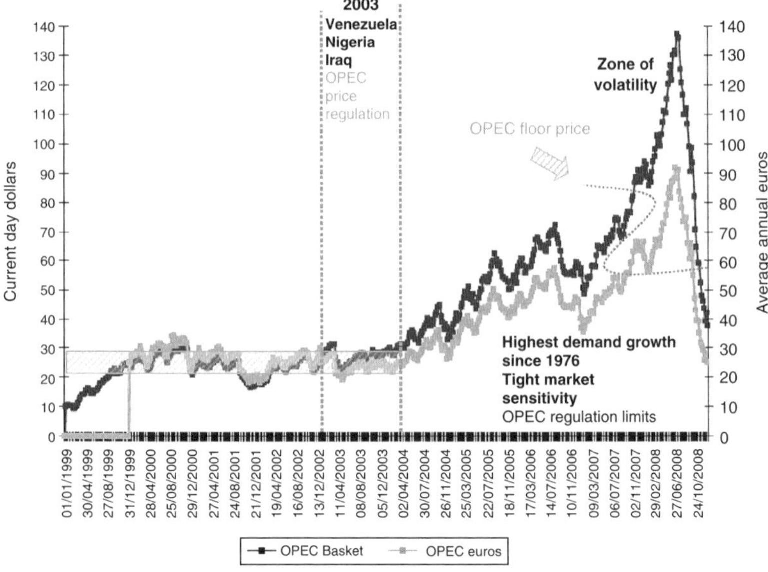

Figure 1.7 Crude oil price, 1999–2008
Source: CGEMP based on data available from Energy Information Administration (2008).

to maintain oil prices within a range of variation with an upper limit of $28 per barrel and a lower limit of $22 per barrel. Price variations were to be controlled by production variations revised periodically. This range of variation was considered as 'a fair pricing': not too low in order to meet the financial requirements of exporting countries (and a minimum profit for high cost producers), not too high to avoid damaging the world economy that had suffered severely after the second oil shock (1979–80). OPEC was very successful in maintaining price stability between 1999 and early 2004 (Figure 1.7), even through the year 2003, which was one of the worst years in oil history with the occurrence of three independent political events:

- In Venezuela, the conflict between President Chavez and the state-owned oil company PDVSA (for the oil money) resulted in a long strike that shut down oil production and oil exports.
- The same year, social unrest in Nigeria (a dispute for oil money) disrupted production and exports.
- Finally, in March 2003, President Bush decided to invade Iraq. Oil production was severely hit.

Despite these events, oil prices did not surge to $100 or $200, because OPEC countries (without Venezuela, Nigeria and Iraq) were able to put on the market the 'missing barrels'. They had at that time considerable spare capacity, which was the key factor in 'controlling' the oil price.

In 2004, the situation changed drastically. There was a surge in oil demand resulting from robust world economic growth and exceptionally high growth in China. However, the role of China in that year 2004 has to be seen in context: China imported about 3 million barrels per day while the United States, stimulated by high economic growth, imported about 13 million barrels per day. The problem was then twofold: production in Venezuela, Iraq and Nigeria had not recovered and the spare capacity of other OPEC countries had vanished. Since 2004, OPEC has lost its power to maintain an upper limit on oil prices. This explains the very high sensitivity of oil prices to any climatic or political event disrupting the balance between supply and demand. However, OPEC is still in a position to maintain a price floor. If the world economic growth is disrupted, if new environmental constraints are imposed, then oil demand could peak and decrease. Then, OPEC would probably regain the market power it had in 1999, cutting production and maintaining a floor price which is no longer at $22 per barrel but more likely to be $80 to $90 (in 2008 dollars). Oil remains, at least partly, a commodity under political control. The oil price peaked on 11 July 2008 at 147 dollars per barrel and then suddenly collapsed to stabilise around 40 dollars at the end of 2008. At the very same moment the global financial crisis hit the real global economy which plunged into a deep recession. What can be expected now is a persisting volatility of oil prices explained by the combination of economic and geopolitical factors.

The evolution of natural gas prices closely follows oil prices with some time lag. Two reasons explain the correlation. First, natural gas and fuel oil are frequently in competition in the heating market. High price differences cannot remain for long. Second is the fact that natural gas supply is frequently ruled by contractual arrangements. In Europe and Asia, the great bulk of gas imports is purchased through long-term contracts. These contracts are not public but each of them contains a price clause with a price formula. Most formulas establish an automatic link between gas price and the price of petroleum products. Sometimes, there are some short disruptions in the parallel evolution of gas and oil prices. They are explained by temporary disruptions of the balance between supply and demand following unusual seasonal temperatures or temporary storage inadequacy.

The surge in coal demand in 2004 was partly driven by higher oil and gas prices as coal became more competitive for power generation. The IEA reference scenario shows that 81 per cent of the increase in coal demand through 2030 will be for power generation. Coal prices should normally be limited by the cost of production in the marginal mines. The range of supply costs varies from $20 per ton (Venezuela, Indonesia) to $50 (United States) (IEA 2006). Actually, the large producers are targeting a price which makes coal slightly competitive with natural gas. Coal prices rose significantly since 2003. Average prices FOB were $30 per ton from 1986 to October 2003, $50 between November 2003 and January 2007 and $55 the forward price between April 2007 and December 2009. New environmental constraints imposed on coal utilisation could impose a limit on coal prices.

Clearly, energy prices are strongly interconnected with a leading role for the crude oil price. The high oil prices of the second oil shock strongly disturbed the world economy. Since 2004, it appears that the gradual and substantial increase in oil prices has had a very limited impact on economic growth in major countries. Economies are fairly insensitive to energy prices. However, higher energy prices are a burden for poor oil importing countries and more generally for the fuel-poor.

The global interconnected economy is vulnerable. Energy resources exist, but continuous economic growth, geopolitical tensions and delays in investment may create scarcity and supply disruptions. National interests and resource nationalism are most often in opposition to the common global goal of building a sustainable energy future. Globalisation tends to aggravate inequalities, and to sustain social and geopolitical tensions. Global capitalism, which partly escapes national regulations, calls for a reinforcement of world regulation, especially for the questions related to climate change, security of energy supply and the financial system. However, actions are limited by inertia and rigidities. Energy prices will adjust the forthcoming disequilibria, and a number of elements tend to indicate that energy prices will be higher and more volatile than in the past.

3 Investments for the future: a balance between private and public initiatives

The future world energy and environmental situation is directly dependent on the investments that will – or will not – be made in the coming years. In the reference scenario of the IEA, and also in the alternative scenario, the need for investment in the energy industry is tremendous. However, the energy sector is full of risks and uncertainties. One of the major uncertainties concerns the policies that will be put in place to protect the environment and limit emissions. Investors and financial markets are taking these risks and uncertainties into account but they know that there are other activities in which they can invest their money. Energy investments are in competition with other investments. Since energy is a political matter, governments may have to intervene in the decision-making process. For oil and gas investments, governments of exporting countries may delay or limit investment on the grounds of resource nationalism or on the grounds of pure economics if they expect a significant increase in the value of their reserves. For power investment, governments may put up some incentives in order to prevent supply disruptions resulting from a lack of investment. Governments may also impose new constraints for increasing energy efficiency, developing renewable sources and limiting all forms of pollution.

3.1 The need for investment

The IEA regularly estimates the investments that are required in the world energy infrastructure. The reference scenario (no change in energy policy) projection of the 2007 *World Energy Outlook* calls for cumulative investment in the energy-supply infrastructure of $22 trillion (in year-2006 dollars) over the period 2006–30. Projected capital expenditures include additional capacity needed to meet demand and the replacement of supply facilities that will be retired during the period. The breakdown shows that 52 per cent of the total goes to the power sector for generation, transmission and distribution. More than half of the whole investment concerns developing countries where demand is growing fast. China alone needs to invest about $2.7 trillion, which represents 12 per cent of the total. In this scenario, which is not sustainable, global energy-related carbon-dioxide emissions increase by 1.7 per cent per year. At the end of the period (2030), the number of people who do not have access to electricity will be roughly the same as that in 2005 – 1.4 billion people – because of the continuing population growth in developing countries.

The reality will probably be different. It will be determined by the parallel decisions of investors and governments.

3.1.1 Investors and governments

Concerning investments in the oil and gas sectors, there are tremendous possibilities in hydrocarbon-rich countries (think of Russia, Iran,

Iraq and Central Asia) but foreign investor ambition is limited by resource nationalism, unilateral changes in contractual conditions (Russia, Venezuela, Bolivia, Kazakhstan) and other limitations. National oil companies' investments are under government control. In open areas (Angola, Congo), competition with the Chinese, the Indian companies and other newcomers is very strong. Production from non-conventional oil is expected to grow substantially, the great bulk of it coming from oil and tar sands in Canada (Alberta). Investment in refining will be concentrated in the Middle East and Asia.

For natural gas, capital needs are highest in North America (OECD) where gas demand is increasing strongly. The rate of investment in the Russian gas industry (one-third of the reserves) is a very important issue for domestic demand, which is growing rapidly, and also for the growing global gas demand.

In the gas sector there are also a great number of projects for pipelines and LNG terminals. A few years ago the world gas market was separated into three different regional markets: North America, Europe (with gas from surrounding countries such as Russia, Algeria, Libya and Nigeria) and Asia, with Japan and Korea importing gas from the Middle East and Southern Asia. Today, the shortage of gas in North America, Chinese and Indian needs, and European willingness to diversify its sources of supply and the rapid development of LNG are leading to a global gas market with increasing opportunities for arbitrage among the three areas.

The development of oil and gas pipelines is also an important matter for energy market integration. The building and the financing of long cross-border pipelines are highly sensitive political problems that imply political negotiations between the countries concerned. Governments must agree to the construction and the operation of the pipeline and also to the fee paid to transit countries. Commercial partners must agree on volume and prices. Natural gas transmission over long distances generally implies a long-term commitment (some 20 years) with respect to volume and price conditions. Sometimes it includes 'take or pay' conditions.[19]

Investment in the power sector, 56 per cent of the total IEA investment projection, is a much more complicated matter. Electricity is a highly political product. Any blackout becomes a political issue. In recent years, a number of serious blackouts have occurred in developed countries: France (1999), California (2001), north-east USA (2003), London (2003), Denmark and Sweden (2003), Italy (2003), Greece (2004), Spain (2004), Germany (2004), Dubai (2005), Los Angeles (2005), Kuwait (2006) and all Western Europe (2006). This list concerns major blackouts where millions of households were hit (Ladoucette 2006). Many blackouts and outages occur regularly in many developing countries.

In the 'good old days', the 'magic of electricity' was brought to households and factories by national or regional monopolies that were most often vertically integrated and state-owned. It was a time of no risks and no competition. Investment decisions were quietly taken through a long-term planning process. At that time, demand forecasting was easier and the investments needed were made in time with a comfortable margin of security to avoid blackouts and outages. There were some disruptions, following storms or accidents, but not comparable to those which have occurred since 2001, the date of the Californian crisis which reflected new changes in the organisation of the power industry.

In the late 1990s, from the United States and the United Kingdom, and later on, from the European Commission came a 'wind of liberalisation' targeting first network industries: telecommunications, electricity, natural gas, railways, airways and postal services. The leading ideas, founded on the theory of contestable markets, were to break the vertically integrated chains, to isolate the segments of the chain that can be considered as natural monopolies and to introduce competition wherever possible.[20] Some markets, which were under monopoly control, became open to new entrants that were expected to bring new competition and innovation. The only remaining monopolies are natural monopolies, a situation in which the control by a single firm is more efficient than competition. The supply of electricity to a given home using two competing wires would be absurd. But, through a single wire, electricity may be supplied by several competing companies.

The major principles of liberalisation of the network industries are: unbundling, third party access, regulation and full market opening.

- *Unbundling* reflects the idea of deconstructing the vertically integrated value chains and separating the competitive activities from the monopoly activities. For electricity, production and supply are competitive activities, while high voltage and low voltage transmission are natural monopolies. For natural gas (in Europe), both high pressure and low pressure gas transmission are considered as natural monopolies.
- *Third party access*: the monopoly segments of the value chains (the wires and the pipes) are essential facilities, meaning that third parties may use the facilities, under certain conditions and with the payment of a fee. Conditions of access must be transparent and non-discriminatory.

- *Regulation*: any monopoly structure may be suspected *per se* – of overpricing, of not investing enough, of being slow in introducing innovation – which is *par excellence* the sting of competition. Therefore, natural monopolies have to be placed under the supervision of a regulator which controls: (i) the tariffs, (ii) the conditions of access, (iii) management efficiency, and (iv) the realisation in due time of the needed investments.
- *Full market opening*: each customer is free to choose its own supplier. In Europe, full market opening for electricity and natural gas has been legally effective since 1 July 2007.

Liberalisation has profoundly changed the traditional organisation of the gas and power industries in many countries in the world. In Europe the engine of the European Community is the building of a single market which is ruled by competition. Obstacles to free circulation of people, goods, capital and services must be removed. Competition is the rule. Monopoly is an exception that has to be clearly justified. The United Kingdom was the first mover in this direction in the early 1990s. The European directives imposing the changes came much later: 1996 for electricity, 1998 for natural gas, 2003 for full market opening. The long discussions of these directives were highly sensitive, from a political point of view, because they called into question the historical organisations that were considered by some nations as being *modèles d'excellence*. Such was the case of France, historically attached to its state-owned vertically integrated monopolies.

Today, the structure of electricity sectors varies considerably over the world. There are still state-controlled vertically integrated monopolies. There are fully unbundled, privatised and competitive systems. There are hybrid models combining state control with some market mechanisms. Many systems are in a transition phase of liberalisation.

The following chapters will return to the power sector issues in the various geographical areas but we would like to make here some general comments on the organisation of the power industry and its relation to the question of investment.

1. In the early 1980s, some economists thought that the liberalisation of the electricity market was a unique occasion to build a market of pure and perfect competition. Electrons are homogeneous goods and the balance between supply and demand has to be instantaneous. Economists thought that, after some measures to lower concentration of the industry, it would have been possible to obtain an atomistic

structure for supply and demand. Clearly, this vision didn't take physics into account. Electricity is a non-storable commodity which circulates at the speed of light (300,000 km per second), and the physical laws governing power transmission prevent the identification of the path followed by the electrons (Kirchhoff's laws). These constraints make the actual functioning of a competitive market more difficult.

2. For years, the core question for electricity markets has been 'What is the best market design?' This is a recurrent question that is not posed for other markets. Many different market designs have been set up. The British, who were pioneers, changed their market organisation several times. The Californian crisis (2001) was 'a failure by design'. For the time being one may say that no market design has proved to be the optimal model. However, the current UK model and the PJM (Pennsylvania, New Jersey, Maryland) organisational structures merit attention.
3. Electricity markets are so complex that they offer a great many opportunities for the exercise of market power. Price manipulation, capacity withdrawal and collusion have been suspected in many places. The temptation of market power is reinforced by an industrial concentration which is frequently quite high with strong dominant positions in various relevant markets.
4. Electricity prices are often determined on power exchanges for several different markets: e.g. the balancing market, day-ahead transactions and futures markets. Depending on the market design, these transactions represent only a fraction of all electricity sales. A large volume of transactions is over-the-counter and through contractual agreements. These contractual prices are freely negotiated. They are not public. In some countries, prices or tariffs are still regulated through state intervention.
5. Investments for the future represent a recurrent and worrying question in all countries that have liberalised their power industry. The problem is as follows: market prices are supposed to be price signals for encouraging investment. In fact they are frequently below the long-run marginal cost, i.e. the cost of producing electricity in a new generating facility. Power systems must be able to meet peak demand with a marginal generating capacity which may be used only a few hours per year. In a competitive environment, there are few incentives to build a peak generating facility. In addition, generators could also be tempted to create scarcity to get higher prices. The problem is aggravated by the fact that liberalisation, and sometimes privatisation, have

reduced capacity margins, meaning that systems are more vulnerable to peak demand. In that case, price volatility might be increased by a combination of adverse factors.

6. The situation which has just been described offers two options: laissez-faire or state intervention. Laissez-faire means that markets and prices will regulate the system, leading to 'boom and bust cycles': high prices will trigger over-investment that will lead to lower prices, under-investment, etc. Intervention means that governments will put in place some sort of capacity mechanism or capacity payment to make sure that the investments needed for the future, including peak capacity, are made at the right moment. The debate is open but, once more, there is no optimal answer to this crucial question.
7. There are other motives for government intervention in the field of power investment: those related to climate change policy which may impose new constraints and restrictions and those that concern other than carbon pollution. Examples are the Large Combustion Plant Directive in the European Union and the Clean Air Act (and the subsequent Clean Air Interstate Rule) in the United States that regulate emissions of sulphur dioxide and nitrogen oxides. Regulation of other pollutants such as mercury and particulates could also play a significant role. More generally, the growing but uneven concern for climate change and pollution will put new constraints and restrictions on the energy investments of the future. Social costs and other externalities have to be taken into account and internalised.

3.2 The decision-making process for energy investment: risk segmentation, risk analysis and risk mitigation

The decision-making process in the energy industry is much more complicated today than it was before globalisation and liberalisation. The change can be explained simply by an environment which is more complex, full of uncertainties and unpredictable. The crucial importance of climate change will trigger the introduction of tougher environmental policies and regulations but the timing is unknown. Climate policy uncertainty enhances investment risk (IEA 2007a).

Not all the energy investors have the same attitude towards risk. International oil companies have always dealt with risk. Investment in exploration is their first risk: one is never sure of finding anything. Political risk is routine: nationalisation, change of contractual conditions, accidents, sabotage. Their risk premium is high. The problem is different for power utilities which, most of the time, have to manage a transition between the old low risk monopolistic structures and the new high risk competitive structures. No other industry has experienced such a radical structural change in such a short period. A new culture of risk has to be built.

The current approach to investment decision in a risky environment is: risk segmentation, risk analysis and risk mitigation. The rationale is still to maximise the expected discounted cash flow (its net present value NPV) but, as the IEA (2007a: 34) puts it elegantly: 'Investment was made if the discounted revenues exceeded the discounted costs. Now, investment is made, only if the discounted revenues exceed the discounted costs by a margin sufficient to overcome the value of waiting.'

Risks related to investment in the energy industry fall into four categories: economic risks, political risks, legal risks and *force majeure* (IEA 2003):

Economic risks apply to the various phases of the project: engineering, construction (a key period where there is no cash flow), operation, maintenance, decommissioning and dismantlement. Risks bear on the expected performance of technologies and equipment, capital expenditures, operating expenditures and the timing of construction. Market risks are related to market uncertainties: such as the price evolution of inputs (such as oil, coal, natural gas and electricity), the price evolution of outputs (such as the price of electricity), the evolution of demand and market capacity to absorb the supply.

Political risks are related to the possible political changes that might take place in the country where the investment is made. They are called country risks. For a given company this is the risk of adverse laws or governmental action hurting companies' interests. Within political risks, *regulatory risk* has a growing importance. It concerns the changes that could be made in the industry structure and industry regulation but also any change in environmental regulation and policy. Once more we come back to the uncertainties related to climate change. The diabolic sequence is very well summarised by the IEA (2007a): 'Uncertainties on GHG emissions create uncertainties in the political responses to increased GHG concentration. Uncertainties in the economic impact of climate change create uncertainties in mitigation policy that will be put in place. All these uncertainties, combined with uncertainties about the cost of abatement technologies, create considerable uncertainties in the financial implications to companies.'

The *legal risks* concern the respect of contractual arrangements, conditions of dispute settlement and litigation. Finally, *force majeure* concerns unexpected events such as wars, terrorist attacks, earthquakes and other natural catastrophes.

Risk segmentation and identification lead to a precise analysis of all risks but with great difficulty in assigning to each risk a probability of occurrence. The main problem is, therefore, to try to mitigate each risk. Risk mitigation is most often ensured through contractual arrangements: such as a contract between the investor and equipment suppliers to guarantee performance, between investors and contractors to guarantee quality and timing, or long-term gas or electricity purchasing contracts to cover price and volume risks. A country's risk can be partly mitigated by national export agencies or by international agencies such as the World Bank's MIGA (Multilateral Investment Guarantee Agency). Governments may play a role, more informal than formal because they are more and more reluctant to provide institutional engagements for the future in periods of high uncertainty. Risk mitigation necessarily implies long and tough negotiations between parties. This is a good and illustrative example of what economists call transaction costs. However, there remain a number of risks that cannot be mitigated.

3.2.1 The case of power investments

Power investments for the future concentrate most of the risks described above. Electricity is an essential good for which there is high political concern with respect to prices and security of supply. Investments have long lead times and face a host of climate change uncertainties. Governmental intervention is almost inevitable for prices, investments, environmental regulation and strategic technological choices (such as nuclear power). Popular opinion is also deeply involved in the opposition to technology choices (nuclear), site implantation and the construction of new high voltage transmission lines. The opposition to new site openings was popularised by the NIMBY concept ('Not In My Back Yard') which is now becoming BANANA ('Build Absolutely Nothing, Anywhere, Near Anybody').

Investment choices for power generation are generally based on the net present value (NPV) of the cost of production per kWh, that is, the discounted capital and operating costs of a new power plant to be built. Various organisations regularly publish the expected cost per kWh produced by coal-fired plants, gas turbines (mostly combined cycle gas turbines), wind farms and nuclear plants.

Today, power utilities are still comparing the expected cost of generation of competing technologies but the choice is complicated by all the uncertainties and risks described above. For a power utility, the future price of carbon is a key question because it is changing the place of coal, gas and nuclear in the merit order. The current environment tends to favour medium-size, rapidly built, less capital-intensive generating units. A combined cycle gas turbine has a low capital cost (but a high fuel cost), a medium size (250 to 350 MW, adaptable step by step to demand evolution) and a high flexibility, i.e. the capability to start or to stop rapidly. The construction takes 18 months compared to seven or eight years for a 1,600 MW nuclear plant which has very limited flexibility and a very high upfront capital cost. But, nuclear power production is domestic (except for the import of uranium) and it produces no CO_2 emissions. This explains why some countries are considering the nuclear option. An interesting example of appropriate risk mitigation for the building of a new nuclear plant is given by Finland (see Chapter 7).

Diversity, with a growing contribution of renewable energy, is still the basic principle for building the low carbon-intensive fuel mix of the future. Diversity makes it possible to experiment with the pros and cons of each energy technology and to measure more precisely the social costs that are associated with each of them. Diversity has value for the society. It also has value in the portfolio mix of a given utility and may appear as a means of risk mitigation. Finally, the investments of the future are highly dependent on each country's strategic options and priorities, but diversity is the key element.

Investments are the key to building a sustainable future. The role of market mechanisms is fundamental but will not automatically lead to sustainability. Social costs and more particularly those related to local and global pollution have to be integrated. New forms of regulation are needed at various levels to make the appropriate and probably urgent corrections to market failures. In the lower carbon-intensive energy balance of the future, diversity is the key element.

4 Challenges for the future

In this chapter, we have brought together the various components of 'the new energy crisis'. We have set the stage for the energy/environment challenges of this century. The global context contains complexities, uncertainties and risks. Among the uncertainties, the most important are those related to the actual effects of climate change, and those related to the economic and geopolitical dynamics of nations. Let us take a look now at the possible evolution of the energy/environment system. We will see more precisely what the challenges are for the future in order to find an answer to the 'equation of Johannesburg'.

4.1 Scenarios for the future

A great number of studies and scenarios, public or private, are devoted to our world energy future. They show a great variety of evolutionary possibilities, from the most optimistic to the most pessimistic.[21] In 2007, the IEA developed three scenarios for world energy development up to 2030: a reference scenario ('business as usual'), an alternative policy scenario with some significant policy changes and a '450 stabilisation case'. Figure 1.8 summarises very clearly the challenges that we are facing between now and 2030. The global energy system was emitting 27 Gt of CO_2 in 2005. In the IEA reference scenario (business as usual) the volume of emissions reaches 42 Gt in 2030. The world scientific community estimates that, in order to avoid an increase in temperature of more than 2°C, we should stabilise emissions at a level of 450 ppm, which means a volume of emissions of 23 Gt in 2030, a reduction of 19 Gt compared with the reference scenario. These figures summarise the new energy crisis. Are we willing to find the 'safe corridor' of evolution? Are we able to do so? The IEA suggests some means for reaching the target: development of carbon capture and sequestration (CCS), development of nuclear and renewable energy, fuel substitution and a great improvement of energy efficiency. This is a challenge for people, governments and companies.

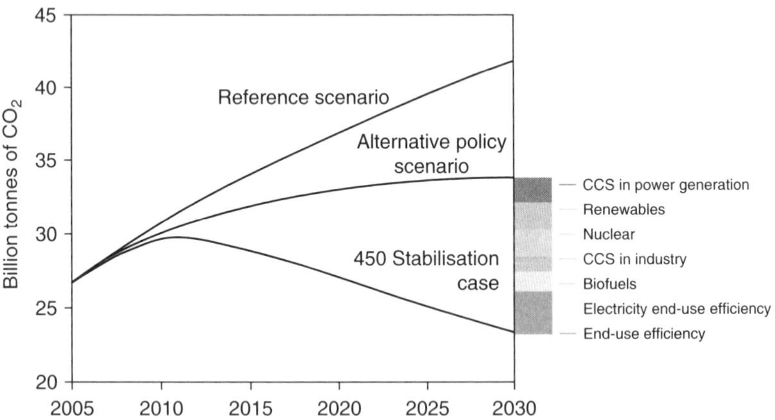

Figure 1.8 CO_2 emissions in the '450 stabilisation case'
Source: CGEMP based on data available from IEA (2007a).

4.1.1 Challenge for people

The international scientific community is largely convinced of the necessity of acting now in order to curb emissions. Outside of this small group, people are more or less convinced that there is a problem but only a minority of them have decided to act, to pay and to put pressure on governments. However, for the time being, the world economy is in the reference scenario. Without greater public awareness and increased pressure on politicians, the world's human community will not resolve the 'equation of Johannesburg'. Awareness means a better scientific knowledge of several key questions: e.g. externalities, climate and clean energy technologies. The precise evaluation of all the social costs that are associated with energy production and consumption is a powerful element for accelerating action. But it takes time.

4.1.2 Challenge for governments

Not all nations and governments have the same sensitivity to environment/energy issues. Some countries deliberately ignore environmental questions while others (Scandinavia was the first) are deeply involved in the battle. European countries share a common political vision of the energy future: lower emissions, improved energy efficiency, more renewable energy and more diversification. Europe has introduced the first market for CO_2 emissions. In the United Kingdom, the government of Tony Blair had a key role in accelerating international public awareness of climate change. He succeeded in persuading George Bush that climate change is an important issue. There is now a change in United States energy policy. The new energy crisis is driving a 'repoliticisation' of energy/environment questions (Helm 2007).

Nations also have a great diversity of attitudes concerning the nuclear question. Some countries still refuse new nuclear plants (Germany, Italy, Austria). Some countries are developing nuclear. Some countries have reactivated the debate. This is the case of the United Kingdom which illustrates the statement of former EU energy commissioner Mrs Loyola de Palacio when she said 'We cannot at the same time reduce significantly GHG emissions and close the door to nuclear.' The case of Finland is another illustration: the country is looking to build a sustainable energy future which includes the building of new nuclear plants. Nuclear will develop but the 'nuclear renaissance' is slow and it does not radically change the energy question. If China builds around 20 new plants before 2020, the share of nuclear in the Chinese energy balance will rise from 1.5 to 2.5 or 3 per cent, but the share of coal will remain about 60 per cent.

Governments' attitudes and actions will finally depend on (i) the pressure of public opinion, (ii) the actual impacts of climate change and (iii) the strength of international dynamics calling for action.

4.1.3 Challenge for companies

The new energy crisis opens, at the same time, new constraints and new business opportunities for the industry. In the rich nations, under growing pressure from public opinion and governments, the industry may expect increasing constraints for reductions in carbon emissions: reductions from their current operation, and in the products that they make and purchase. More and more companies are being invited to define and measure their 'carbon footprint', a concept that indicates the level of emissions the company is responsible for. These types of constraints are distorting global competition. They illustrate a major opposition, especially in Europe, between competitiveness and the reduction of emissions.

On a more positive note, the connection between energy and climate change offers a wide range of new opportunities: clean technologies, energy services and energy efficiency. The ebullient innovation process which is taking place in the automotive industry provides an illustration. The private sector is the primary driver for the demand for project-based credits, e.g. the Clean Development Mechanism (CDM) and Joint Implementation (JI), the two most important market instruments associated with the Kyoto protocol.[22]

Financial markets are now taking into account companies' attitudes towards climate change and environmental responsibilities. In a rather cynical way, the financial community is less interested in scientific evidence than in companies' responses: 'Forget the science. It's time to focus on threats and opportunities.'

5 Conclusion

In this chapter we have explained the challenges arising from the new energy crisis. The following chapters will analyse how the main geographical areas of the world are going to act and react and how the planet will be able to resolve the 'equation of Johannesburg'.

We have to keep in mind that the geopolitics of the planet is changing. Today, the European Union, Russia and North America account for 16 per cent of the world population. By the middle of the century, in 2050, they will account for 11 per cent of the 9.2 billion inhabitants, 8 billion of them being located in Asia, Africa and Latin America.

All the factors that we have described in this chapter, coming from economic, geopolitical and cultural contexts are now moving within the dialectical uncertainties of the future. Recalling what we wrote in the introduction, the resolution of the equation will come through the combination of three factors: actions from governments, institutions and companies – adaptations which could be costly and painful for the most vulnerable – and a final variable of adjustment: prices.

Notes

1. The role of the Intergovernmental Panel on Climate Change (IPCC) is to assess the scientific, technical and socio-economic information relevant to the understanding of the risk of human-induced climate change.
2. In the statistics of the International Energy Agency (IEA), the world total primary energy demand, which is equivalent to total primary energy supply, represents the total demand including power generation, other transformations, own use and losses. Primary energy refers to energy in its initial form. Primary energy is transformed into secondary energy mainly in refineries, power stations and heat plants. The total primary energy demand includes traditional biomass and waste such as fuel wood, charcoal, dung and crop residues, some of which are not traded commercially.
3. The study has selected seven fields to measure the negative impact that power generation may have: human health (mortality and morbidity), construction materials, crops, effects on global warming, sound environment and the state of the ecosystems.
4. Kenworthy (2003).
5. Energy independence as measured by the ratio of domestic primary energy production to primary energy consumption.
6. CERA (Cambridge Energy Research Associates) is a well-known energy consulting firm based in Cambridge, Massachusetts.
7. On the possible effects of climate change and economic consequences see: Stern (2006); IPCC (2007), IMF (2008).
8. *Oil and Gas Journal* and other sources. Some reserve valuations are still made under the Securities and Exchange Commission's rules in which disclosure definitions are based upon technologies of the 1970s.
9. The rate of recovery indicates the amount of oil which can be recovered in a given field. The average rate of recovery was 25 per cent in 1973. It is now 30–35 per cent and could be significantly improved in certain fields.
10. The 'peakists' have created an association called ASPO (Association for the Study of Peak Oil).
11. CERA's argumentation is developed in *Why the 'Peak Oil' Theory Falls Down: Myths, Legends and the Future of Oil Resources*. Peter M. Jackson, November 2006.
12. Royal Norwegian Ministry of Finance: *Petroleum Industry in Norwegian Society 1973–1974*. Parliament Report 25.
13. The Muslim world represents about 1.2 billion people. The most important countries, in terms of Islamic population are Indonesia (185 million), Pakistan (140 million), India (124 million) and Bangladesh (116 million).

The largest countries where the majority of the population is Muslim are Iran (65 million Muslims), Turkey (65 million), Egypt (61 million), Algeria (30 million) and Morocco (28 million). *Libération*, 28 August 2007.

14. The 'seven sisters' were Exxon, Mobil, Shell, BP, Gulf Oil, Texaco and Standard Oil of California (Chevron). They had a dominant position in the international oil market and were accused of being a cartel by the Federal Trade Commission which issued a report in 1952: *The International Oil Cartel.*
15. Clean Development Mechanism: a project-based mechanism allowing Annex 1 countries to invest in carbon-reduction initiatives in Non-Annex I countries. The Annexe 1 country must have a commitment inscribed in Annex B of the Kyoto protocol and must have ratified the protocol (see special Box in Chapter 9).
16. Statistics on coal generally include steam coal (for boilers and power generation), coking coal (for the steel industry), brown coal and peat. In 2005, steam coal represented about 70 per cent of global coal production.
17. Some entrepreneurs of the early American capitalism in the nineteenth century were called the 'robber barons'.
18. Samuel Huntington (1996).
19. In a 'take or pay' contract the buyer has the obligation to take each year the quantity contracted for and, even if he doesn't take it, he has to pay for the entire contracted volume. TOP can be rephrased as 'Take, and even if you don't take, you have to pay.' This very hard clause can be explained by the fact that the sunk cost of building the gas line is very high and bankers want a guarantee of a regular cash flow.
20. Baumol et al. (1982).
21. Among the available public scenarios, the most frequently quoted are those of the IEA, IPCC, the World Energy Council and Shell.
22. Clean Development Mechanism: see the Box in Chapter 9. Joint Implementation: a project-based mechanism whereby two Annex 1 countries share carbon reduction credits. To qualify, both countries must have a commitment inscribed in Annex B of the Koyoto protocol and must have ratified the protocol.

References

Baumol, W. J., Panzar, J. C. and Willig, R. D. (1982) *Contestable Markets and the Theory of Industry Structure*. New York: Harcourt Brace.

Boissieu, C. de (2006) *Division par quatre des émissions de gaz à effet de serre de la France à l'horizon 2050*. Rapport du groupe préside par Christian de Boissieu. La documentation française.

Chevalier, J. M. (1975) *The New Oil Stakes*. Harmondsworth: Penguin Books.

Chevalier, J. M. (2004) *Les grandes batailles de l'énergie*. Paris: Gallimard.

Chevalier, J. M. (2006) 'Security of Energy Supply for the European Union', *European Review of Energy Markets*, 1, 3.

Dessus, B., Laponche, B. and Le Treut, H. (2008) 'Rechauffement climatique, l'importance du méthane', *La Recherche*, March.

European Commission (2003) *External Costs: Research Results on Socio-Environmental Damages Due to Electricity and Transport*. Directorate General for Research.

Friedman, T. (2006) *The World is Flat: a Brief History of the Twenty-First Century*. New York: Farrar, Straus & Giroux.
Fukuyama, F. (1992) *The End of History and the Last Man*. New York: Free Press.
Giraud, P. N. and Lefèvre B., 'Les défis énergétiques de la croissance urbaine au sud. Le couple "Transport–Urbanisme" au coeur des dynamiques urbaines', *Regards sur la Terre*, Rapport Annuel AFD-IDDRI, Presses de la Fondation Nationale des Sciences Politiques.
Helm, D. (2007) *The New Energy Paradigm*. Oxford: Oxford University Press.
Huntington S. (1996) *The Clash of Civilisations and the Remaking of World Order*. New York: Simon & Schuster.
IEA (2003) *World Energy Investments Outlook*. Paris: IEA.
IEA (2006) *World Energy Outlook*. Paris: IEA.
IEA (2007a) *Climate Policy Uncertainty and Investment Risk*. Paris: IEA.
IEA (2007b) *World Energy Outlook: China and India Insights*. Paris: IEA.
International Monetary Fund (IMF) (2008) 'The Fiscal Implications of Climate Change', Washington.
IPCC (2007) *Fourth Report: Climate Change 2007*.
Jackson, P. M. (2006) *Why the 'Peak Oil' Theory Falls Down: Myths, Legends and the Future of Oil Resources*. Cambridge, MA: CERA.
Kenworthy, J. (2003) 'Transport Energy Use and Greenhouse Gases in Urban Passenger Transport Systems: a Study of 84 Global Cities', Working Paper, Murdoch University, Perth.
Keppler, J. (2007) 'Security of Energy Supply: a European Perspective', *European Review of Energy Markets*, 2, 5.
Ladoucette, V. de (2006) *Energy Security in a Fragile World*. CERA Special Report, Chatham House, November.
Lorenzi, J. H. (ed.) (2008) *La guerre des capitalismes aura lieu*. Paris: Perrin.
Mandil, C. (2008) *Energy Security and the European Union: Proposals for the French Presidency*. Report to the French Prime Minister.
OECD, Nuclear Energy Agency and International Atomic Energy Agency (2008) *Uranium 2007: Resources, Production and Demand*. Paris: OECD.
Platts Nuclear Publications (2007–9).
Stern, N. (2006) *The Economics of Climate Change* (Stern Review). Cambridge: Cambridge University Press.
World Bank (2008) 'Global Monitoring Report 2008: MDGs and Climate Change – Accelerating and Sustaining Development', *Global Monitoring Report*. World Bank Publications.
Yergin, D. (1991) *The Prize: the Epic Quest for Oil, Money and Power*. Oxford: Blackwell.
Yergin, D. and Stanislaw, J. (1998) *The Commanding Heights: the Battle Between Governments and the Marketplace that is Remaking the Modern World*. New York: Simon & Schuster.
Yergin, D. (2006) 'Ensuring Energy Security', *Foreign Affairs*, 85, 2: 69–82.
Yergin, D. (2008) 'Oil has Reached a Turning Point', *Financial Times*, 27 May.

2
The Questioned Sustainability of the Carbon-Dependent Asian Dynamics

Patrice Geoffron and Stéphane Rouhier

1 Introduction

Asia is that part of the world where progress in economic growth is at its fastest pace and the place where the most people (close to 1 billion) lack access to modern forms of energy. This combination means that, in the coming decades, a huge increase of energy consumption is anticipated: this economic 'success story' will lead to higher energy needs to feed the world's factory, which Asia is in the process of becoming; in the meantime, the urbanisation associated with this process and the improvement of living standards will also contribute to keep energy demand under pressure. Considering the size of the economic area and this catching-up process, the energy developments in Asia will obviously be the main engine of the future evolution of the global energy system.

Considering the wide diversity of Asian national situations, a specific focus will be made on three countries: China, India and Japan.[1] These economies are today major actors on the energy scene, being ranked, for example, second, third and fifth respectively in terms of electricity consumption. In the future, due to their skyrocketing growth, China and India might be responsible for 45 per cent of the increase in world energy demand by 2030 (IEA 2007a). Japan is also interesting since this country is both highly dependent in terms of oil and gas imports but is probably the most efficient country in terms of energy use.

The chapter's aims will be twofold:

- Presenting and analysing the position of this area in global markets and the dynamics of energy consumption in China, Japan and India and highlighting the economic and geopolitical issues related to the future energy choices inside this economic area and the related impact outside of it (global energy markets, pollution, security of supply, etc.).

- Examining the available options to overcome those problems, especially by underlining the possible cooperation or competition among the countries of this region.

2 Main economic and geopolitical issues arising from the Asian energy surge

2.1 Consumption and production trends and prospects[2]

The Chinese energy mix can be explained by the willingness of the Chinese authorities to remain as independent as possible. Coal represents more than 60 per cent of the primary energy production and may stay high in the future, due to the abundance of local resources in China (second at the world level); this high dependency will not be substantially modified in the future. Oil represents 19 per cent of the energy mix today; demand has long been met by domestic oil, but, with the recent boost in demand, the country is now a net importer. The third main energy source in China is traditional biomass. Although its use has decreased thanks to electrification, it still represented 13 per cent in 2005. Up to 2030, according to the IEA reference scenario, the main phenomenon to be anticipated is that, for the three sources of carbon, China will become a first rank importer with 13.1 mb/d of oil, 128 bcm of natural gas and 95 million tonnes of coal (IEA 2007a). Despite the important local resources, China will be responsible for 7 per cent of the international trade in coal. The explanation essentially relates to the increased needs for electricity generation (more than 1,300 GW), widely dominated by coal.

The Indian energy mix is also characterised by strong coal use as India benefits from huge reserves. Indeed, in 2005 coal represented 39 per cent of the energy mix, and is mainly used for power generation purposes. Like China, the two other main energy sources are oil (24 per cent) and traditional biomass (29 per cent) used in rural areas. Even though gas use has recently increased, it only represents 5 per cent of the energy mix. By 2030, coal use is to rise to 47 per cent of the energy mix also driven by the increased need for electricity. The oil share will grow to 25 per cent while the other sources of energy, as in China, will increase their shares (8 per cent for natural gas, for instance) at the expense of traditional biomass (15 per cent). India is in roughly the same position as China, but with smaller resources. It has also very limited oil and natural gas that respectively amount to only 0.4 per cent and 0.6 per cent of world reserves. Recent discoveries will enable natural gas production to grow until 2030 and, as coal is the most favoured energy option in India, its production will increase by a factor of almost three by 2030. For natural gas, imports are projected to be multiplied by eight by 2030.

The Japanese energy mix has substantially changed over the last fifty years. In the 1950s, coal supplied half of Japanese energy demand and a third was supplied by hydropower. Then, the government decided to use oil extensively but, since the shocks in the 1970s, it was decided to diversify the energy mix and to use energy more efficiently. In 2005, 47 per cent of the energy came from oil, 21 per cent from coal, 15 per cent from nuclear and 14 per cent from natural gas. In terms of future perspectives, as Japan is quite aware of the environmental constraints, notably global warming, oil and coal will see their share reduced to 36 per cent and 19 per cent respectively; while nuclear and natural gas will see theirs increase to 21 per cent and 18 per cent by 2030. Japan is even less well endowed with natural resources than its two counterparts. Indeed, it has no significant domestic resources of fossil fuels except coal. Thus, it has always relied almost entirely on imports for natural gas and oil (coal being costly). This is, along with environmental concerns, one of the reasons for the Japanese decision to dynamise their nuclear energy policy; nuclear power will represent almost a quarter of the energy demand in 2030.

2.2 Global repercussions in hydrocarbon energy markets

In any of the scenarios proposed by the IEA, China and India will import more than half of the gas needed by 2030. India will rely almost entirely on the Middle East while China will import its gas from several suppliers, mainly Australia, other developing Asian countries and transition economies. Because of their increasing weight in terms of gas consumption, Asian countries are likely to have an increasing impact on gas prices, in the medium term. Indeed, non-OECD Asia, led by China and India, is forecasted to be the fastest growing region in terms of natural gas consumption and should become a net importer by 2020 which would increase Asian reliance on external resources (US EIA 2007).

In terms of global oil use, China and India will continue to put pressure on the global demand as their combined share is likely to increase from 12 per cent in 2005 to 20 per cent in 2030, according to the IEA reference scenario, and developing Asia will increase its share of global oil consumption from 18 per cent to 28 per cent (IEA 2007a). Therefore, oil demand will mainly be driven by this region, as it represents 54 per cent of the additional demand over the period 2005–30. It should also be noted that Japan and Korea already have a 100 per cent oil-import dependence and that developing Asia's ratio is likely to increase to over 70 per cent in 2030. Consequently, along with this increase in oil consumption, the region will have to face a higher dependency on imports. This is likely to strongly affect the global market since this fastest growing region in the world relies heavily on external resources.

Coal is the dominant energy source in China and India which represent 45 per cent of world coal use and will account for three-quarters of the additional demand over the period 2005–30. The other Asian countries are also extremely active in the global coal market. For instance, Japan, South Korea and Taiwan are the three largest coal importers[3] in the world and India and China are ranked fifth and eighth respectively (Koizumi and Maekawa 2007). In total, Asian countries represent more than half of the world's imports and developing Asia will account for 65 per cent of world consumption in 2030. On the exporting side of the market, Asian countries represent only 11 per cent of the coal exported worldwide and China became a net importer of coal in 2007 (IEA 2007a). Therefore, a rise in coal demand will strongly affect the global market as Asian countries will probably rely on supply from other regions. In the past few years, Asian countries have viewed coal as a cheap and available fuel and thus as a way of developing their electricity generation. As a result, coal prices skyrocketed between 2000 and 2005, from approximately US$38 to US$63 per ton of steam coal. 2004 can be seen as the turning point when total coal demand increased by 8.5 per cent and this was following a 7.5 per cent increase in 2003. In terms of prices, this led to a spectacular 48.6 per cent rise measured by customs unit values for the two major IEA importing areas (EU-15 and Japan) (IEA 2005). Prices have also gone up because producing countries have not committed themselves to investments early enough. This event is referred to as the '2004 coal crisis' (Koizumi and Maekawa 2007).

Coal prices are likely to remain high in the medium term. Indeed, while the IEA forecasts a small decline in price up to 2010 before rising again to the 2005 levels, the Institute of Energy Economics Japan (2006) predicts that prices will remain at the same level. In both cases, supply is expected to smoothly meet demand. Indeed, after the 2004 crisis, producing countries have upgraded existing mines, developed new ones and invested in transportation infrastructure. Therefore, coal production is likely to increase in the future. This explains the IEA prediction in which supply will exceed demand, resulting in lower prices. However, Chinese and Indian demand for coal may rise even faster than predicted and thus could lead to other crises in the years to come.

So, the '2004 crisis' is interesting in the sense that it may prefigure a continuing pressure on prices in Asian countries in the future as they will increase their share of total world demand and may account for 87 per cent of the additional demand over the period 2005–30.

Box 2.1 The feedback of high energy prices on growth in Asian countries

Among oil-importing countries, the poorer the countries are, the greater the reduction in GDP will be due to high oil prices (UNDP/ESMAP 2005). The IMF estimated in 2004 that a sustained US$10 oil-price increase would lead to a reduction of 1.5 per cent in the GDP in those countries after one year. In Asia, this would induce an overall reduction of the GDP of 0.8 per cent after a year with some countries suffering more, such as the Philippines which could lose 1.6 per cent of its GDP (IEA 2004). One of the adjustments available to reduce the effects of an oil-price increase relates to the price elasticity. Indeed, in non-OECD countries this price elasticity tends to be much lower in absolute value than in OECD countries, lowering the reduction in consumption that follows such a price increase (UNDP/ESMAP 2005). What should also be taken into account is the domestic pricing policy. When oil prices increase, these governments face a dilemma. Either they continue to subsidise the energy and help the consumers, which results in increasing the drain on public finance, or they let prices rise to partly reflect world prices. Since oil products tend to affect a large part of the population, few governments have been willing to use the second, unpopular, solution. (Note that fuel protests demanding greater fuel subsidies took place in June 2008 in Malaysia, Thailand and Indonesia.) However, the drain on the treasury has strong implications for possible developments in education, infrastructure improvement, and so on. Another effect of a price increase that occurs in poor, developing countries is the rise in traditional biomass consumption, through fuel switching, which increases indoor air pollution as well as deforestation. Regarding China, Kahrl and Roland-Holst (2008: 657) propose an interesting focus on exports, with three main conclusions: 'First, gross exports are significantly more cost vulnerable to energy prices than other final demand activities because they are, by assumption, more import-intensive. Gross exports are more than 40 per cent more oil cost vulnerable than gross household consumption, indicating that a rise in oil prices would raise the cost of producing goods and services for export by significantly more than the cost increases of goods and services for household consumption. This suggests that, among China's final demand activities, exports would be most affected by rising global and regional energy prices... Second, across final demand activities and for both net and gross final demands, the Chinese economy is uniformly more cost vulnerable to oil than coal. In other words, a one percent increase in the price of oil has a much larger effect on input costs than a one percent price increase in the price of coal. Though perhaps somewhat counterintuitive given that the coal is the Chinese economy's primary energy input, this apparent dichotomy reflects oil's significantly higher cost per unit energy vis-à-vis coal... Third, given that oil costs embedded in gross exports are the most energy cost vulnerable activity in China's economy, and that a significant portion of these costs are incurred outside of China's borders, there are likely limits to what China's central government can do to shield the economy from energy price volatility in the near term.'

2.3 Environmental impact

To get a general idea regarding the current environmental situation of the Asian countries, one can refer to the Environmental Performance Index from the Yale Center for Environmental Law and Policy (2006).[4] Of the 133 countries ranked for 2006, Asian countries show the poorest performances. Apart from Malaysia, Japan and South Korea, the other ASEAN+3 countries[5] are all ranked beyond 50th place. One can also add that Mongolia, Tajikistan, India, Yemen, Bangladesh and Pakistan are all between the 115th and the 130th place.

2.3.1 At the global level

As most of the Asian countries are in the process of development, they are not responsible for most of what is happening now. Indeed, while China and India account for 8 per cent and 2 per cent respectively of the cumulative CO_2 emissions over the period 1900–2005, the US and the EU are responsible for more than half of these emissions (IEA 2007a).

Nevertheless, in the past few years, Chinese and Indian emissions have soared. Thus, according to certain sources,[6] China could already be the first CO_2 emitter worldwide and Japan and India are already fourth and fifth. And this Asian predominance is likely to increase as China and India account for 56 per cent of the increase in CO_2 emissions between 2005 and 2030, in the reference scenario of the IEA, so that they would be the first and third CO_2 emitters in 2030, while Japan would be the fifth. At that time, Chinese emissions are forecasted to be 66 per cent higher than those of the US, ranked second. Therefore, challenges in terms of CO_2 mitigation are huge and should, at some point, take place in Asia. This point will be further discussed in the next section.

However, currently, climate change is not the form of pollution that worries most Asian inhabitants. Indeed, countries like India or China (as well as most of the South East Asian countries) rely heavily on coal and thus suffer from air pollution at both local and regional levels.

2.3.2 At regional and local levels

Regional pollution is caused by acid rain which occurs when SO_2 and NO_X are mixed together in the air. This leads to the creation of acidic compounds that are absorbed by clouds, which in turn makes rain or snow more acidic. This impacts on vegetation, soil, crop yields, buildings and public health. It not only affects the place where the pollution is emitted but it can be transported over thousands of miles. For example, Japan and Korea are suffering from Chinese pollution through acid rain, while Bangladesh suffers from Indian pollution. As for the rankings, China is already, by far, the first SO_2 emitter in the world and forecasts show that both Chinese and Indian emissions of NO_X and SO_2 will rise steadily. For example, Indian emissions could double between 2005 and 2030.

Measuring the impact of acid rain appears to be difficult as most of the studies differ widely in their conclusions. In any case, the figures shown are not encouraging. Indeed, Chang and Hu (1996) found that the average yield for vegetables in Chongqing (China) has been reduced by 24.5 per cent. Another study undertaken by Zhang and Wen (2000) showed that Chinese agricultural production has already been lowered by between 5 and 10 per cent by acid deposition. Lastly, a World Bank study (2007a) showed that crop losses in China due to SO_2 and acid rain represented 30 billion RMB in 2003.

The reasons for local pollution are the same: a heavy reliance on coal and non-conventional biomass that both emit large quantities of noxious gases (carbon dioxides, sulphur doxides, nitrous dioxides, particulate matter, etc.). Indoor air pollution occurs mainly in poor areas as it is related to the use of traditional biomass. For example, in developing countries, people tend to rely on wood, dung or crop residues for domestic energy.

Exposure to this polluted air leads, among other things, to respiratory illness, cancer, tuberculosis, low birth weight and eye disease. For

Table 2.1 Sulphur dioxide concentration in selected cities

Country	*Cities*	*Sulphur dioxide (in micrograms per cubic metre, 1995–2001)*
China	Guiyang	424
China	Chongqing	340
China	Taiyuan	211
Iran	Teheran	209
China	Zibo	198
China	Qingdao	190
China	Jinan	132
Brazil	Rio de Janeiro	129
Turkey	Istanbul	120
China	Anshan	115
Russian Fed.	Moscow	109
China	Lanzhou	102
China	Liupanshui	102
Japan	Yokohama	100
China	Chenyang	99

Source: World Bank (2007b).

example, exposure to biomass smoke may explain 59 per cent of rural cases and 23 per cent of urban cases of tuberculosis in India. In China and India, it has been shown that two-thirds of women with lung cancer were non-smokers (Bruce et al. 2000). And according to Zhang and Smith (2007), indoor air pollution is responsible for more than 400,000 premature deaths annually in China.

As for outdoor air pollution, the burden of disease is mainly shared among developing countries, and Asia alone represents 65 per cent of that global burden related to outdoor air pollution (Cohen et al. 2005). For example, in the last ranking of the world's most polluted cities, China accounted for twenty of them[7] and according to a World Health Organization report (2004), only 31 per cent of Chinese cities meet WHO standards in terms of air quality. Table 2.1 underlines the fact that the majority of the cities with high SO_2 concentration are in Asia, and mostly in China.

There is, therefore, a high cost to be paid for that pollution as it causes illness and death. In its last report on this topic, the World Bank (2007a) assessed the economic cost of Chinese pollution at between 3 and 6 per cent of GDP in 2003, depending on the methodology used.

3 Available strategies to tackle these issues

In this section, we will review strategies available that are either currently used or that could be implemented to overcome those problems. China's and India's increasing weight in international trade imposes an obligation to contribute to enhancing global energy security, an effort that will also impact the rest of the world. First, the security of supply will be tackled and then we will analyse what can be done to improve the state of the environment. These two issues should be considered together by the countries concerned, as most of the solutions overlap. Indeed, the policies needed to improve the environment also partly improve the security of supply, and vice-versa. They are both part of a so-called sustainable energy future that needs to be devised for Asian countries.

3.1 Strategies to improve security of energy supply

The concept of security of energy supply means, basically, the capacity of any given country to have access to adequate, affordable and reliable supplies of energy. This security may be 'challenged' by energy market instabilities, technical failures or physical security threats (Chevalier and Keppler 2007; IEA 2007c). As far as the Asian governments are concerned, the biggest concern is oil. Indeed, gas does not currently represent a very large share of their energy mixes and coal comes mainly from indigenous production and its resources are better shared. Therefore, as we shall see, the strategies of Chinese, Indian and even Japanese governments are focusing on oil resources.

3.1.1 Energy efficiency

The most common way to measure energy conservation is to use energy intensity which is the amount of energy used to produce one dollar of GDP. Developing Asian countries are not performing very well and there is, in this respect, a good deal of room for improvement. On the other hand, Japan is one of the best countries in this regard. In recent years, China and India have enacted laws to increase their energy savings. For example, China has set intensity targets for its provinces, has closed inefficient power plants and heavy industry and has reduced its subsidies to better reflect international prices, to increase energy efficiency and to make producers take into account their effect on the environment.

In 2004, China's medium- and long-term plan for energy conservation, which gives specific targets for a number of sectors (industrial, transport, building), was released. It provided guidance to achieve the required

20 per cent improvement in energy efficiency needed between 2005 and 2010 by the Eleventh Five-Year Plan, although the government still needs to invest more in that area (Brookings Institution 2006a). Lastly, the Top 1,000 Enterprises Energy Efficiency Programme is aimed at setting targets for the largest Chinese firms in terms of energy intensity and the goal of the Energy Conservation in Government Programme, in force since 2006, is to improve the energy efficiency of government institutions (IEA database).[8]

In India, a Bureau of Energy Efficiency was set up in 2001 to coordinate policies and programmes. It has also decided to introduce compulsory labels in terms of energy efficiency in buildings and to improve existing power plants and networks so that losses are reduced. The Indian Tenth Five-Year Plan set an objective of energy savings of 13 per cent through efficiency improvements, without specifying the expected efficiency gains per sector. Efforts concerning conservation and efficiency have already paid off. Indeed, petrol and diesel consumption increases have slowed thanks to better roads and vehicles (Brookings Institution 2006b). Even Japan, which is one of the best performing countries, has set a target of reducing its energy consumption per unit of GDP by 30 per cent by 2030 (Toichi 2006).

The Asian Development Bank (2006) decided to launch a project on energy efficiency improvement that will in the future enable the implementation of lending and non-lending assistance programmes. In the past, the Asian Development Bank has already financed a number of projects such as $150 million for the Industrial Development Bank of India for large industrial energy efficiency improvement projects (ADB 2006).

3.1.2 Diversification of energy resources

Currently, most of the Asian developing countries are heavily reliant on coal. For example, it represents 47 per cent of the energy mix[9] in developing Asian countries; while in Japan 47 per cent of its energy is derived from oil. Diversifying would reduce reliance on one or the other of these energy sources and thus they would be less affected in case of an energy crisis.

Projects of this kind are currently undertaken in many countries. For example, China is developing its natural gas market both through the construction of liquefied natural gas (LNG) terminals and pipelines and is supporting, as is India, its renewables market. Renewables and nuclear power represent domestic sources of energy and, thus, enable a country to be less reliant on others. Renewable energy sources are also

Table 2.2 Expected growth of renewables in China under the 2004 law

Sector	*Current capacity (end 2004) (MW)*	*Expected capacity in 2020 (MW)*
Wind power	560	20,000
Solar energy	50	1000
Biomass	2000	20,000
Hydro power	7000–8000	31,000

Source: Asif and Muneer (2007).

widely distributed and can, therefore, reduce transmission losses and costs for electricity generation. Projects of hydropower, nuclear power, biomass, biofuels, wind and solar generation have been launched in these two countries. Table 2.2 shows current and expected renewables capacity if China is to meet the 2004 Renewable Energy Promotion Law's expectations, that is to say annually develop new and renewable energy sources to amount, by 2015, to 2 per cent of the country's total energy consumption (Asif and Muneer 2007).

It should be noted that India was, in 2004, the fifth-largest wind energy producer in the world. The Indian Ministry of Petroleum and Natural Gas's Hydrocarbon Vision 2025 report specified that natural gas should become the main Indian source of energy. As for diversification of the energy mix, Japan has set an objective to reduce to less than 40 per cent the share of oil in its energy mix and to increase to 30–40 per cent at least the share of electricity produced by nuclear power (Toichi 2006).

In terms of cooperation, it can be noted that India has been assisting Bhutan and Nepal both technically and financially in the development of their hydropower resources. More generally, China has cooperation projects in the fields of renewables with the Asian Development Bank, the World Bank, the US, Italy, Germany and Britain, while India has renewables projects with UNDP/GEF, the World Bank and the Asian Development Bank. These two countries are also establishing cooperation in the field of renewables through academic committees, forums and manpower sharing (Liming 2007).

3.1.3 Diversification of oil supply sources and routes

The rationale behind oil routes diversification is to reduce the chances of piracy and terrorism. Currently, two-thirds of Indian oil comes from the Middle East and most of the additional oil needed by India up to 2030 will be supplied by this region because it has huge resources and because it is close to India (IEA 2007a). As for China, the Middle East currently accounts for 45 per cent of Chinese imports while the remaining 55 per cent comes from Africa, the former Soviet Union and other developing countries. The additional oil needed by 2030 is likely to be supplied by the Middle East and the former USSR countries. Once again, these two regions have both the resources and proximity as advantages for supplying China.

The main problem of this reliance on the Middle East and Africa is the fact that oil is shipped through two critical shipping channels, namely the straits of Hormuz and Malacca. The Strait of Malacca is 900 kilometres long with a flow of 12 million barrels per day in 2006 (Masuda 2007). At its narrowest point it is only 500 metres wide. The Strait of Hormuz, which lies at the mouth of the Persian Gulf, is the busiest oil-shipping route in the world with a flow of 13.4 million barrels per day corresponding to 16 per cent of the global oil supply (and close to one-third of trade volume). Growing tensions have highlighted the potential risks of piracy and terrorism and, more generally, the risk of supply disruptions. Therefore, countries have recently tried to bypass these two routes.

To do so, China and India have tried to agree on the building of pipelines. Oil can be transported by pipeline, rail, road or ships. This partly explains the Chinese willingness to obtain oil from Africa and the former Soviet Union. The other reason is related to the diversification of oil suppliers. Indeed, it makes it possible to reduce the share of imports coming from the Middle East. This is a strong motivation for the Chinese authorities as they fear a potential oil blockade in case of a conflict over Chinese Taipei. Consequently, the idea is to reduce imports from the Middle East and by sea.

There have, therefore, been many pipeline projects involving various Asian countries. This illustrates how cooperation can lead to mutual benefits as countries get either energy or markets for their resources. One example is the Kazakhstan–China pipeline, whose eastern leg was completed in 2005 and for which discussions are underway to increase capacity. There is also under discussion a project called Iran–Pakistan–India (IPI) that would transport gas from Iran to India and Pakistan. This pipeline would provide benefits to the three countries: Iran would find markets for its gas, India and Pakistan would obtain the natural gas they need and Pakistan would receive transit fees. Other projects for pipelines such as Turkmenistan–Afghanistan–Pakistan–India and Myanmar–Bangladesh–India are also currently under discussion (Brookings Institution 2006b).

These initiatives underline the Indian objective of pushing forward gas as the energy source of the future. The East Siberian Pacific Ocean (ESPO) pipeline is a good example of both cooperation and competition among Asian countries. The idea is to transport oil from Russia to China and Japan. The question was which leg would be constructed first: either a pipeline from East Siberia to the Russian Pacific coast or a spur to China. After some indecision, the Chinese won the political battle over the Japanese and therefore they will be served first, in 2009, with a capacity of approximately 0.6 million barrels per day (IEA 2007a).

3.1.4 Equity oil and overseas acquisitions

In the late 1980s, China launched its policy of 'going-out' which corresponds to acquiring equity stakes in exploration and production assets overseas. This policy has two goals, the first being to supply its increasing oil needs and the second, to create competitive international companies. Concerning energy security, the rationale is to enable the country to establish its oil reserves. Therefore, in case of a supply disruption from any of the country's oil suppliers or in case of an oil blockade, it would be able to reroute the physical flow of its oil. In addition, it provides a hedge in case of price increases. Indeed, if prices were to rise, the government could cap the price and divert the oil to the national market.

Today, Chinese national oil companies own 0.6 million barrels per day of oil production overseas while India owns 0.1 million barrels per day. Chinese companies have invested in countries such as Kazakhstan, Sudan, Indonesia, Nigeria and Angola. The volume currently controlled by Asian national oil companies is rather small compared to their needs and is mostly sold on the market rather than shipped towards the home country. On the other hand, between 40 and 50 per cent of the Chinese companies' oil is shipped to China (IEA 2007a).

These acquisitions of equity oil abroad have set the stage for fierce competition among Asian countries. Recently, Chinese national oil companies won four deals over Indian companies: in 2004 in Angola, in 2005 in Kazakhstan and Ecuador and in 2006 in Nigeria (Paik et al. 2007). Japanese companies are also investing abroad. Currently, the proportion of oil produced by Japanese national oil companies corresponds to just under 15 per cent of the oil consumed in 2005 and the government wishes to increase this share to 40 per cent by 2030, a figure that might be difficult to reach (Niquet 2007).

This policy is not supported by a strong consensus. For example, the IEA is rather dubious of the effectiveness of such a policy and is in favour of a market solution. According to the IEA, transparent international oil markets coupled with energy efficiency measures and oil stocks would achieve better results than a countervailing policy of building up market power. Besides, quantities held by Asian national oil companies are not significant enough to really improve their physical oil supply and thus, to protect them against a price hike.

3.1.5 Strategic oil reserves

Since the 1970s, strategic oil reserves have been recognised as crucial to limit the effects of oil supply disruptions. For example, the International Energy Agency requires its member states to have the equivalent of, at least, 90 days of net oil imports in oil reserves. The ASEAN Petroleum Security Agreement (APSA) signed in Manila in 1986 is an example of cooperation concerning oil. Indeed, this agreement mentions an ASEAN Emergency Petroleum Sharing scheme. This means that if at least one member country is in critical shortage, the ASEAN oil exporting countries will help the affected state(s). This scheme has not been used yet.

The imperative for stockpiling has been of increasing concern in recent years because of the fear of supply disruptions (e.g. 9/11, instability of the Middle East, OPEC spare capacity decrease). Of the Asian countries, Japan and South Korea have reserves of at least 90 days of net imports as they are both OECD and IEA members. Currently Japan reports a 160-day oil stockpile, including state-owned reserves (90 days) and private reserves (70 days) (Toichi 2006). Non-OECD countries also hold stockpiles. Examples are Thailand (mandated by its 1978 Fuel Act) and the Philippines (Giragosian 2004). It should also be noted that the Chinese government launched a project of strategic oil storage construction. Four sites are being built with a capacity of almost 100 million barrels, which corresponds to 27 days of current imports and should be completed by 2008. Second and third phases of construction with two additional sites of 200 million barrels are planned. This would raise the capacity to the equivalent of 75 days of 2015 net imports. The Indian government is doing the same and a strategic petroleum reserve was due to start in 2007 and should be completed by 2012. It will have a capacity of 36 million barrels, the equivalent of 19 days of current net imports, and is expected to increase to 110 million barrels (IEA 2007a).

Box 2.2 Focus on nuclear issues in Asia

Japan decided in 1973 that nuclear energy was a strategic priority, because of concern about security of energy supply. This decision was reinforced later on by considerations about climate change. Japan now has 55 power reactors in operation providing some 30 per cent of its total electricity. Two reactors are under construction and 12 other units are firmly planned. The government's scenarios call for 60 GWe producing 40 per cent of electricity in 2017 and 90 GWe providing some 60 per cent of electricity in 2050 (plus 20 GW thermal of nuclear capacity for hydrogen production). Japan started by importing nuclear plants from Great Britain and the United States, but very soon organised its own nuclear industry under US licence and gradually became more independent. Today, Japan is the only country to have facilities for a complete nuclear fuel cycle without being a nuclear weapon state and Japanese manufacturers are world leaders. Toshiba owns a majority stake in Westinghouse, Hitachi has a joint nuclear venture with General Electric and Mitsubishi cooperates with Areva. Japanese industry and government are developing 'Generation 3++' designs of BWR and PWR for deployment in 2020, with the objective to reduce the capital and kilowatt-hour costs by 20 per cent and to extend lifetime to 80 years. Today, Japan has the lead among countries in the international R&D partnership, Generation IV International Forum, for development of sodium-cooled fast reactors (SFR). Thus, Japan's nuclear energy development benefited from a long-term view and continuity in scientific and political decision-making bodies. As a weak point, we can mention a complex decision-making process – the need to forge consensus of many national and regional institutions – leading to a rather average capacity factor for operating nuclear power plants. For example, a large part of the country's nuclear capacity was shut down for over one year after the discovery in 2001 of a cover-up of inspection results. All seven units at Tokyo Electric Power Co.'s Kashiwazaki-Kariwa site remained off line in July 2008, a year after a severe earthquake revealed that design assumptions for ground acceleration were wrong. Nevertheless, beside the steady development of nuclear energy in Japan, Japanese industry is now in the strongest position to participate in world development of nuclear energy.

South Korea began developing nuclear energy at the end of the 1960s. The first nuclear power plant, imported from the United States, operated in 1977. Its electric power industry followed soon, ordering reactors from the US, from France and Canada. Then the Koreans decided to develop their own nuclear design and engineering industry, the Combustion Engineering (now Westinghouse) System 80 pressurised water reactor under licence. In 2008, 20 power reactors totalling 17.5 GWe were operating, providing 40 per cent of the country's total electricity supply. The overall plan calls for 27 GWe in operation by 2020, providing 45 per cent of total electricity, and some further increase to about 60 per cent of total electricity by 2035. Korean industry is now ready and eager to export its products. There are still some licensing issues, especially regarding the US or Chinese market. However, the Koreans have an agreement with Indonesia to export four reactors by 2016 and the Koreans are looking toward Vietnam and Thailand as potential customers.

Korea also has a large research and development programme in nuclear technology: fast breeder reactors, high-temperature reactors for hydrogen production, and smaller plants for desalinisation of water and electricity production.

China first developed its nuclear energy towards weapons production, serious development of civil nuclear energy only starting at the end of the 1980s. The Chinese developed their own power plant design of 300 megawatts electric, but very quickly turned to foreign companies (Canada, France and Russia). In 2007, the Chinese ordered four AP1000 reactors from Toshiba/Westinghouse and two EPR reactors from Areva NP, all with 'Generation 3' designs. The Chinese AP1000s are the first of their kind in the world. The Chinese have also developed with French technology transfer a 0.6-GWe design reactor and have built two units of that size. In 2007, Chinese industry representatives indicated that they will develop a Chinese Generation 3 reactor which should be cheaper than imports. Presently, the cost of an installed kilowatt is estimated at US$2,000 for foreign reactors. In 2020, according to the goal determined by the central government, nuclear share in total electricity should represent 5 per cent (close to 60 GWe). For 2030, expectations are around 160 GWe of nuclear in operation. The new policy is to approve all the regional electricity companies' plans provided, if they are based on 'reasonable' criteria. China also has basic technologies for uranium enrichment and fuel reprocessing coming from its military programme. But for commercial enrichment, China has imported Russian gas centrifuges, and is considering a commercial reprocessing plant using French technology. Thus, China's rapid expansion of electricity – 260 GW of new plant, mostly coal, by 2020 – combined with increasing awareness of the detrimental effect of harmful emissions and problems with transport of coal from northern coal fields to southern consumption centres, as well as the greater availability of capital for large investments, make possible a very strong expansion of nuclear electricity – perhaps even larger than is presently considered. The limiting factor, at least for the next five to ten years, could be the availability of specialised personnel.

The situation of nuclear energy in **India** is very strongly influenced by the country's decision not to sign the Nuclear Non-proliferation Treaty and to develop a nuclear weapons capability. India exploded its first nuclear device in 1974 and was practically excluded from any international cooperation or commercial exchange in the nuclear field. India developed its own nuclear reactor technology, based initially on Canadian natural uranium-fuelled pressurised heavy water reactors. It now operates a fleet of small reactors (0.2 GWe) and more recently, larger PHWRs (0.5 GWe). The current nuclear generating capacity is around 4 GWe out of India's 110 GWe total electric generating capacity. India has developed a complete national industry, including production of heavy water, uranium mining and refining, fuel fabrication and reprocessing of spent nuclear fuel. However, its industry has some quantitative limitations, notably for uranium production. The Indian nuclear reactors in 2008 operated at only about half of nominal power because of insufficient fuel supply. India's initial nuclear energy strategy was based on three stages: natural uranium-fuelled PHWRs, producing plutonium; fast breeder reactors, producing plutonium and uranium 233 from thorium; and heavy water thorium-plutonium or thorium-U233 reactors. This strategy was dictated by India's international isolation and by its limited uranium resources but with large thorium resources. The second stage, fast breeder reactors, is under development. A 40-MW(thermal) test reactor (practically a copy of France's Rapsodie, which began operation in 1967) has operated successfully from 1985 up to now. A 0.5 GWe prototype fast reactor, of technology similar to France's Phenix/Superphenix, is under construction and scheduled for operation in 2010. Four other similar fast breeder reactors are planned to operate before 2020. However, this strategy allows only a rather slow development of nuclear power production. Therefore, India has negotiated an agreement (approved in September 2008) to apply international safeguards without joining the NPT. With such agreement, India expects to expand its nuclear power generating capacity by importing light water reactor technology and enriched uranium, in addition to building more pressurised heavy water reactors of domestic design. The optimistic plan is to have 40 GWe in 2020 and at least 25 per cent of total electricity supplied by nuclear in 2050.

C. Pierre Zaleski (CGEMP).
Source: IGCAR (2008), WNA, Platts Nuclear (2008).

3.2 Strategies to lower the environmental impact

3.2.1 Market improvements

In Asia, households without access to electricity use non-conventional biomass, whose effects on public health and the environment have already been mentioned. Hence, reducing energy poverty would have a positive effect on the environment by reducing the use of biomass: when a household has access to electricity, its first use is for lighting, and thus it reduces the use of kerosene or biomass. The Indian government has recently started tackling the problem of providing improved electricity access to the entire country. In 2005, the electrification rate was 62 per cent, though the number of people using biomass as cooking and heating fuel has increased in the past few years. The Electricity Act of 2003 forces Indian utilities companies to supply electricity everywhere, including to villages.

But, electricity does not simply replace biomass and a unique and single transition process does not exist. Each country or even region has its own procedures. To shift from traditional biomass to modern energy three components are needed: availability, affordability and cultural preference. Indeed, biomass is often seen as free and readily available, and even if it is bought it will probably be cheaper than any of the other energy sources. Concerning traditions, Indian households, even those that are rich, still use their biomass stoves to prepare their traditional bread. Therefore, strategies other than the electrification of the country must be pursued, such as upgrading kitchen ventilation and the efficiency of biomass cooking stoves in poor households. Indeed, whereas biomass cooking stoves using dung offer an 8 per cent energy efficiency and 9 per cent using wood fuel, coal and charcoal cooking stoves have 25 per cent energy efficiency and those using natural gas, kerosene or LPG reach 50–60 per cent (IEA 2007a).

In China, electrification has been a great success and the electrification rate reached 99 per cent in 2005.[10] In the 1980s the Chinese government took supportive measures with the creation, for example, of local firms and basic infrastructure. This has helped to alleviate poverty and increase human welfare. However, rural households still use biomass for cooking purposes as clean fuels are not affordable and widespread in rural areas and thus still constitutes a heavy environmental burden. Therefore, there is also a need for action in China, through off-grid renewables and solar and biogas thermal technologies that are being promoted by the Eleventh Five-Year Plan.

Historically, fossil fuels such as oil and coal have been heavily subsidised. This has distorted the market and increased the quantity of oil and, above all, coal consumed. Reducing subsidies makes it possible to reduce the incentive to consume polluting sources of energy. In any case, subsidies have not always achieved their primary goals. For example, the subsidies on kerosene and liquefied petroleum gas (LPG) in India were intended for the poor, in order to make them consume less biomass. However, it has been shown that 40 per cent of the subsidies benefited the richest 7 per cent of the population (IEA 2007a). The Committee on Pricing and Taxation of Petroleum Products announced that limiting this scheme only to households living under the poverty line would cut down by 40 per cent the quantity of subsidised kerosene consumed.

Other policies are needed in India to enable a change from biomass to cleaner cooking fuels in the poorest households and new schemes are being implemented. One example is the Deepam LPG scheme that subsidises the technology but not the fuel. Indeed, the state government of Andhra Pradesh provides a free LPG connection but does not offer any subsidy for fuel refill (IEA 2007a).

Lastly, it should be noted that as developed countries, both South Korea and Japan have binding commitments concerning carbon emissions, thanks to the Kyoto Protocol. Therefore, 'new' market instruments such as joint implementation, clean development mechanisms and the carbon trading scheme will be available for them to reduce their emissions (see Chapter 9).

3.2.2 *Technology improvement*

As electricity generation is one of the greatest emissions culprits, technology will play a central role in environmental preservation. Indeed, clean coal technologies will be able to reduce emissions of carbon dioxide, sulphur dioxide, nitrogen dioxide and dust. Carbon capture and storage (CCS), now available, can currently capture 85 per cent of the CO_2 that would otherwise have been emitted. However, they reduce power plant efficiency by 8–12 per cent. The current price of such a technology is quite high but it is expected to decrease to less than $25 per tonne of CO_2 by 2030 (IEA 2007a).

New technologies will be able to increase thermal efficiency. By using critical or super-critical coal-fired power plants, Asian countries would significantly increase their efficiency and thus reduce their emissions.[11] Therefore, this issue is extremely important in China and India as they are and will continue to be heavy coal users. According to the IEA (2007a), if Chinese and Indian coal power plants were to reach the efficiency of OECD plants that would provide, in 2030, a decrease in CO_2 emissions of 650 Mt, which would represent 2 per cent of the global emissions of that year. These two types of technologies (super-critical power plants and CCS) can be coupled and will have a great role to play in combating global warming.

Figure 2.1 shows current Indian and Chinese coal-fired power plants' efficiency and emissions as well as OECD, state-of-the-art and R&D plants. One can see the gains that would be achieved by upgrading current Chinese and Indian coal-fired power plants to the level of OECD plants, for example, both in terms of CO_2 emissions and efficiency.

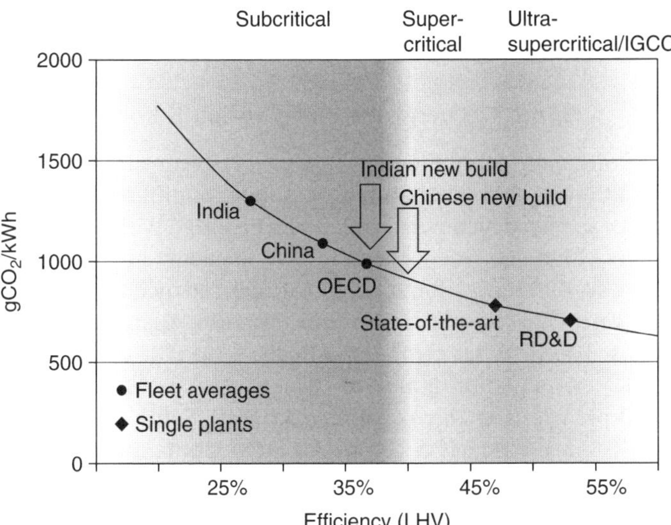

Figure 2.1 CO_2 emissions from coal-fired power plants
Source: International Energy Agency (2006c).

International cooperation will be more and more required to facilitate the deployment of such technologies in developed as well as developing countries. For example, in 2006, the Asia-Pacific Partnership on Clean Development and Climate was launched by Australia, Japan, South Korea, the United States, China and India. Further cooperation takes place through the Asia-Pacific Economic Cooperation and the East Asia summit. There is also bilateral cooperation between the United States and China and India where carbon sequestration and the use of clean coal technologies in the power sector are becoming important. The European Union has also launched programmes with China for a more sustainable use of energy and with India focusing on clean coal technologies and clean development mechanisms (CDM). In addition to this kind of cooperation, China and India are the largest markets for CDM projects. China alone represents half of the CDM projects undertaken worldwide and has developed expertise in identifying and designing CDM projects. A law on 'Measures for Operation and Management of the Clean Development Mechanisms Projects' has also been ratified to set priorities and establish general conditions. As for India, it is already the second CDM market and 75 per cent of total savings take place in energy-related projects (IEA 2007a). Therefore, China and India will be the stage for technological improvements either through cooperation or with CDM projects.

Box 2.3 Energy insecurity in South Korea

South Korea's situation of energy insecurity is very similar to that of Japan, although there are naturally a number of differences. The two countries share a dependence which remains high with respect to oil and a very high dependence with respect to energy imports. The following two differences stand out most: dependence on the Middle East is somewhat lower in South Korea than in Japan (80.7 per cent as opposed to 87.9 per cent, respectively, in 2005); conversely, the South Korean situation is notably worse in terms of energy intensity and per capita energy consumption. First, South Korea's energy insecurity means that it is competing with China and with Japan, but without being able to wield either China's demographic and economic weighting or Japan's level of development, despite the fact that the Koreans have been trying hard for many years to equal and even surpass Japan – something which they have achieved in a number of heavy industries. The size and per capita income difference vis-à-vis its neighbours means that, to a certain degree, South Korea is caught in a stranglehold. Secondly, as with Japan, the will to diversify its imports from the Middle East is driving South Korea to develop ties with Russia and, to a lesser degree, with Central Asia. If collaboration with Russia to secure oil from Siberia and Sakhalin-1 and gas from Kovykta and Sakhalin-2 is primordial in Japan's case, it is no less so for South Korea, with the particularity that – since it is a continental country – the eventual links by oil pipeline and gas pipeline would appear to be easier. However, it should be recalled that the situation in North Korea is an added factor which could complicate this relationship, or which could rather act to facilitate it, in view of the fact that a definitive solution to the energy problems of North Korea must involve the creation of supply channels throughout the peninsula. Third, the need for more regional cooperation on matters of energy security is even more pressing for South Korea than for Japan, due to questions of size and those linked to the necessary improvement of inter-Korean relations. Thus, Seoul has a special interest in promoting energy cooperation in North-East Asia. Lastly, it is worth noting that the efforts to promote renewable forms of energy have been somewhat greater in South Korea than in Japan. While Seoul has set an official target to attain a 5 per cent contribution of these energies (not including hydroelectric power) by 2011, Japan has no official target of this type, despite having a higher per capita income, and notwithstanding its technical superiority in certain industries.

Source: Based on Bustelo (2008).

4 Conclusion

Until recently, industrialised countries have dominated world energy use. The situation in Asia demonstrates that this situation is currently changing: industrialisation, increasing living standards and population growth tend to increase energy consumption in developed countries, with subsequent consequences for global sustainability. The transition of Asian economies, in various degrees, from low efficiency solid fuels to oil, gas and electric power and from agriculture to industrialisation as well as from low to high motorisation is thus a collective issue.

China is at the very centre of this process, with primary energy consumption that has exceeded domestic energy production since 1994, leading to a substantial expansion in oil imports. In the longer term, with the maturity of the economy, production is moving towards less energy-intensive activities and more energy-efficient technologies are being introduced. The Asian model of economic growth is, therefore, at risk with an urgent task for authorities to establish long-term energy policies to avoid excessive shocks to economic development. Moreover, we must keep in mind that the spectacular Asian economic growth has led to unprecedented environmental consequences.

Notes

1. Chapter 4 presents a wider view of the other Asian countries through the lens of poverty issues.
2. The data are taken from the IEA's *World Energy Outlook* (2007a) and its 'Reference Scenario'.
3. This calculation has been made excluding lignite.
4. http://research.yale.edu/envirocenter/
5. Namely, in that ranking, the Philippines, Thailand, Indonesia, Myanmar, China, Vietnam, Laos and Cambodia (ranked by EPI results from the best to the poorest).
6. Netherlands Environmental Agency (2007), http://www.earthtimes.org/articles/show/74352.html
7. World Bank: China quick facts.
8. IEA Energy Efficiency Policies and Measures database.
9. This figure is due to the dominant size of China and India in the region. Other countries such as Pakistan, Bhutan and Nepal are not necessarily reliant on coal.
10. However, it must be kept in mind that a household that can only light a single bulb is as connected as one that can run many electrical appliances.
11. For the moment, in China, only 6.5 per cent of all the coal-fired power plants are super-critical and in India projects are still in the planning stage (IEA 2007a).

References

Asian Development Bank (2006) *Report on the Energy Efficiency Initiative*.

Asif, M. and Muneer, T. (2007) 'Energy Supply, its Demand and Security Issues for Developed and Emerging Countries', *Renewable and Sustainable Energy Reviews*, 11: 1388–1413.

Brookings Institution (2006a) *China*. Energy Security Series, Brookings Foreign Policy Studies.

Brookings Institution (2006b) *India*. Energy Security Series, Brookings Foreign Policy Studies.

Bruce, N., Perez-Padilla, R. and Albalak, R. (2000) 'Indoor Air Pollution in Developing Countries: a Major Environmental and Public Health Challenge', *Bulletin of the World Health Organization*, 78, 9: 1078–92.

Bustelo. P. (2008) 'Energy Security with a High External Dependence: the Strategies of Japan and South Korea', MPRA Paper 8323, University Library of Munich, Germany.

Chevalier, J. M. and Keppler, J. H. (2007) 'Security of Energy Supply: a European Perspective', *European Review of Energy Markets*, 2, 2: 5–37.

Chang, Y. and Hu, T. (1996) 'The Economic Cost of Damage from Atmospheric Pollution in Chongqing', manuscript, Chongqing Environmental Protection Bureau.

Cohen, A. et al. (2005) 'The Global Burden of Disease Due to Outdoor Air Pollution', *Journal of Toxicology and Environmental Health Part A*, 68, 13–14: 1301–7.

Cruz, R. V. et al. (2007) 'Asia Climate Change 2007: Impacts, Adaptation and Vulnerability', contribution of Working Group II to the Fourth Assessment Report of the Intergovernmental Panel on Climate Change, in M. L. Parry et al. (eds), *Climate Change 2007*. Cambridge: Cambridge University Press, pp. 469–506.

Day, K. (2005) *China's Environment and the Challenge of Sustainable Development*. Armonk, NY and London: M. E. Sharpe.

Giragosian, R. (2004) *Energy Security in East Asia*. Institute for the Analysis of Global Security.

HM Treasury (2006), Stern Review on the economics of climate change.

Institute of Energy Economics Japan (2006) *Asia: World Energy Outlook 2006*. IEEJ.

International Energy Agency (2004) *Analysis of the Impact of Higher Oil Prices on the Global Economy*. IEA.

International Energy Agency (2005) *Coal Information 2005*. IEA.

International Energy Agency (2006a) *Coal Information 2006*. IEA.

International Energy Agency (2006b) *World Energy Outlook 2006*. IEA.

International Energy Agency (2006c) *Focus on Clean Coal*. IEA.

International Energy Agency (2007a) *World Energy Outlook 2007*. IEA.

International Energy Agency (2007b) *Energy Security and Climate Policy: Assessing Interactions*. IEA.

International Energy Agency (2007c) *Contribution of Renewables to Energy Security*. IEA information paper.

International Energy Agency (2007d) *Key World Energy Statistics 2007*. IEA.

Kahrl, F. and Roland-Holst, D. (2008) 'Energy and Exports in China', *China Economic Review*, 19, 4, December: 649–58.

Koizumi, K. and Maekawa, K. (2007) *Impact of Changes in Indian Coal Supply/ Demand Outlook on International Coal Markets*. Institute of Energy Economics Japan.

Lee, C.-C. and Chang, C. P. (2008) 'Energy Consumption and Economic Growth in Asian Economies: a More Comprehensive Analysis Using Panel Data', *Resource and Energy Economics*, 30: 50–65.

Liming, H. (2007) 'A Study of China–India Cooperation in the Renewable Energy Field', *Renewable and Sustainable Energy Reviews*, 11: 1739–57.

Masuda, T. (2007) 'Geopolitics of Oil and Gas Pipelines', paper presented at the University Paris-Dauphine, CGEMP.

Morikawa, T. (2006) *Natural Gas and LNG Supply/Demand Trends in Asia Pacific and Atlantic Markets*. Institute of Energy Economics Japan.

Niquet, V. (2007) 'Energy Challenges in Asia', Les notes de l'IFRI.

Paik, K. W. et al. (2007) 'Trends in Asian NOC Investments Abroad', background paper, Chatham House.

Platts Nuclear Publications (2007–9).

Population Reference Bureau (2007) '2007 World Population: Data Sheet'. PRB.

Smith, K. R. (1999) 'The National Burden of Disease from Indoor Air Pollution in India', in G. Raw, C. Aizlewood and P. Warren (eds), *Indoor Air 99*, the 8th International Conference on Indoor Air Quality and Climate, August. Edinburgh and London: Construction Research Ltd., pp. 13–18.

Srivastava, L. and Misra, N. (2007) 'Promoting Regional Energy Co-operation in South Asia', *Energy Policy*, 35: 3360–8.

Streifel, S. (2006) 'Impact of China and India on Global Commodity Markets, Focus on Metals, Minerals and Petroleum', draft working paper, Development Prospects Group, World Bank, Washington, DC.

Toichi, T. (2006) *International Energy Security and Japan's Strategy*. Institute of Energy Economics Japan.

United Nations Development Programme/World Bank Energy Sector Management Assistance Programme (2005) *The Impact of Higher Oil Prices on Low Income Countries and on the Poor*. UNDP/ESMAP.

US Energy Information Administration (2007) *International Energy Outlook 2007*. US Department of Energy.

Villar, J. A. and Joutz, F. L. (2006) *The Relationship between Crude Oil and Natural Gas Prices*. US Energy Information Administration, Office of Oil and Gas.

Wang, J. et al. (2001) 'The Appraisal of Social and Economic Impacts of Acid Rain and Sulphur Dioxide Control Programs in the Total Control Area in the Tenth Five-Year Plan'. Environmental Policy Research Series.

World Bank (1997) *China 2020: Clear Water, Blue Skies*. World Bank.

World Bank (2007a) *Cost of Pollution in China: Economic Estimates of Physical Damage*. World Bank.

World Bank (2007b) *World Development Indicators 2007*. World Bank.

World Health Organization (2004) *Environmental Health Country Profile: China*. Geneva: WHO.

World Nuclear Association. Information Papers (2008).

Yale Center for Environmental Law and Policy (2006) *Pilot 2006 Environmental Performance Index*. New Haven: Yale Center for Environmental Law and Policy.

Yang, J. and Schreifels, J. (2003) 'Implementing SO_2 Emissions in China', OECD Global Forum on Sustainable Development: Emissions Trading. Paris: OECD.

Zaleski, C. P. (2006) 'The Future of Nuclear Power in France, the EU and the World for the Next Quarter-Century', *Perspectives on Energy*.

Zhang, J. and Smith, K. R. (2007) 'Household Air Pollution from Coal and Biomass Fuels in China: Measurements, Health Impacts, and Interventions', *Environmental Health Perspectives*, 115, 6: 848–55.

Zhang, K. and Wen, Z. (2000) 'Resource Efficiency: Resource Diversification and Realizing Sustainable Development', Shanghai Environmental Sciences, February: 47–50.

3

Russia and the Caspian Region: Between East and West

Nadia Campaner and Askar Gubaidullin

At the turn of the twentieth century, the Russian Empire emerged as the world's largest crude oil producer and exporter with the rapid development of the oil industry in the Baku region.[1] In Baku, a cosmopolitan city where East meets West, entrepreneurs of various national origins made fortunes out of oil while eminent scientists such as Dmitry Mendeleev brought their expertise to the needs of the oil industry. After a tumultuous century of ups and downs, modern Russia is back again on the international scene as a major oil and gas exporter, manifesting ambitions to become an energy superpower in the context of surging global demand for energy. New Russia's strategy highlights that its unique Eurasian location and vast hydrocarbon resources should be able to ensure the security of supply to both its Western and Eastern neighbours in the twenty-first century. The same ambitions are cherished by the former Soviet republics around the Caspian Sea. Rich in oil and gas, they are emerging as new important suppliers of energy resources. Their full potential is still untapped.

1 Russian fuel and the energy sector

Russia is one of the largest energy producers in the world: it is the first producer of natural gas, the second for oil, and the fourth largest electricity producer after the USA, China and Japan. Unlike most other industrialised nations, Russia is self-sufficient in energy. Russia has a wide natural resource base, which includes the world's largest natural gas reserves and the second largest coal reserves. It also has an important hydroelectric potential, as the country possesses about 9 per cent of the world's hydro resources. The Russian power sector has a relatively diversified fuel mix. Conventional thermal power plants account for 66 per cent of the electricity generation, hydropower for 18 per cent and nuclear for 16 per cent.[2]

Russia produces and consumes a lot of energy. It is the second largest consumer of gas after the United States. Natural gas currently covers about half of its energy needs. Its primary energy consumption per capita is comparable to that of the EU-15 or Japan (about 4.5 Mtoe/capita/year), but it is less that of the Nordic countries or North America. The Russian economy is characterised by its particularly high-energy intensity (about three times higher than that in Europe or Japan). At the same time, its electricity consumption per person is lower compared to other industrial countries. Note that this major oil producing country has a rather low oil consumption level per capita (0.9 tonnes). Russia ranks third for carbon dioxide emissions, after the US and China, with twice less CO_2 emissions per head than the US.

The specificity of Russia is that most of its population and industries are located in the European part of the country while the natural resources are found in the Asian part, thousands of kilometres away from the main consumption centres. This implies high transportation costs and requires bringing in manpower to those inhospitable regions. Production costs are also considerably higher than in Saudi Arabia or Qatar for instance. Moreover, there are wide variations in consumption levels across regions, with affluent Moscow and the industrial Urals already experiencing acute deficits of electricity supply.

The fuel and energy complex – TEK in Russian[3] – remains a key sector of Russia's contemporary economic development. The TEK provides a large share of the country's industrial output (about a quarter of the GDP) and yields the biggest tax revenue (one-third of the federal budget in 2006). Electricity and energy consumption per capita is generally correlated with living standards. But energy is much more than a base and a locomotive of Russia's economy: it is a vital sector that enables life in this vast and cold country. Any electricity blackout or shutdown in the boilers on a cold winter day could prove catastrophic: cuts to hot-water supply mean that residential water pipes and apartment radiators may simply freeze and burst, making the whole area uninhabitable. Those unfortunate residents of Siberia or Sakhalin that experience winter electricity cut offs certainly learn at first hand what energy is.

Russia is the largest net exporter of energy with oil and natural gas dominating its external trade. In fact, hydrocarbons represent over half of the external trade, a proportion that has increased significantly over the last decade. Approximately two-thirds of all the crude oil and one third of the natural gas produced in Russia are exported. The bulk of the

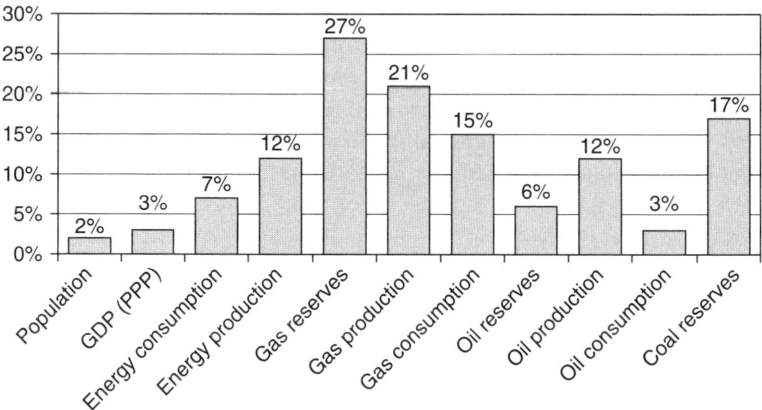

Figure 3.1 Russia, the world and energy: main indicators (% share)

crude oil and gas exports are delivered to EU countries and Russia is by far Europe's biggest external gas supplier. In the future, Russia may also boost exports of oil and gas to Eastern markets, such as China, Japan and India, if new transport infrastructures are built. Russia also exports electricity, mostly to the ex-USSR countries, and it is the third largest exporter of coal which represents about 10 per cent of global coal exports. The share of raw materials increased from 42 per cent in 1995 to 65 per cent of the total exports value.[4] Even though it is expected to decline in the long term, it is likely to remain a core sector and its natural resources wealth remains Russia's only 'competitive advantage'.

With only 2 per cent of the world population, 3 per cent of the world's GDP (Figure 3.1) and the economy dependent upon the trade of raw materials, modern Russia may not regain its superpower status. However, in the current context of growing demand and limited supply, energy has become not only the most important strategic commodity but also a powerful geopolitical tool. Energy is a means for Russia to restore its lost influence and to play an increasing role on the international scene.

2 The legacy of the past

In 1920 Russia was in ruins after the First World War and the civil war. Consequently the Bolshevik government adopted an ambitious long-term plan for the recovery and development of the national economy, based on the accelerated development of the energy sector and heavy industry. The State Commission, headed by a brilliant electrical engineer,

Gleb Krzyzanowsky, elaborated the overall plan for the electrification of Russia (GOELRO Plan). According to this plan, electricity production would almost quintuple compared to pre-war figures. The English science fiction novelist H. G. Wells, who visited Soviet Russia in 1920, called this plan the 'Utopia of the electricians'. He wrote: 'Can one imagine a more courageous project in a vast flat land of forests and illiterate peasants, with no water power, with no technical skill available, and with trade and industry at the last gasp?'[5] Yet, the GOELRO Plan turned out to be a true success story: by 1935 Russia had become the third largest electricity producer after the US and Germany, and was transformed into a major world industrial power by the late 1930s. Whole new industries were created from scratch in a short period of time with limited foreign aid and capital.[6] Massive investment in technical education and research assured the development of skills and technology. However, the human cost of the crash industrialisation programme was high: the heavy industry and the power generation sectors were literally built on the sweat and blood of millions of Soviet people. The success of the programme in the USSR in the context of the great American depression led the Soviet authorities to claim the supremacy of the socialist central planning economy over market capitalism. From 1929 the entire Soviet economy was managed on the basis of the Five-Year Plan. The liberal tendencies of Lenin's New Economic Policy (NEP) were abolished. The long-term central planning principles in the energy sector were applied in various forms both in developing countries (China, India) and in Western Europe (notably in France) after the Second World War.

One of the weaknesses of the USSR's economy was its over-dependency on two single sources of fossil fuel supply: 80 per cent of its coal was extracted in the Donetsk basin (most of it is in Ukraine) and over 75 per cent of the oil came from the Baku region (Azerbaijan). During the Second World War, German troops occupied the Donetsk Basin and were determined to capture the oil fields of Baku. It was of paramount importance to Hitler to secure fuel supplies for his war machine. In 1943 the Soviet army stopped the Nazis in a decisive battle at Stalingrad which was to become a turning point in the Second World War. During the war period, many Soviet industries had been evacuated further inland to the Urals, which resulted in a major shift of the industry eastwards. The coal and iron ore deposits in the Kuznetsk Basin pushed the development of new industrial centres in south-western Siberia. As for oil, it was largely due to the theoretical works of Ivan Gubkin, a great oil geologist,[7] that immense resources were discovered between 1930 and 1940 in the Volga–Ural region.

Thanks to the development of the Volga–Ural region after the Second World War, the Soviet Union boosted its oil output and by the early 1960s it had become the second largest oil producer in the world. During the war exports stopped but they were resumed by the mid-1950s. However, the arrival of cheap Soviet oil on the international market triggered concerns among the Western majors. At that time, Moscow used an aggressive dumping price strategy in order to gain market share, forcing the 'Seven Sisters' to cut their posted prices for Middle East crude.[8]

In 1957, Andrei Trofimuk,[9] a prominent petroleum geologist, decided to move to a newly opened research centre in Novosibirsk. Asked what an oilman could do in Siberia where no one had seen much oil, he replied: 'Siberia is literally floating on oil.' A few years later, a huge fountain of Siberian oil announced the birth of a new era of oil and gas exploration. It was a field geologist from Baku, Farman Salmanov, who took a risk and ventured to explore the virgin territories of the Ob river basin. Western Siberia soon became the new 'oil and gas frontier', gradually replacing maturing oil fields in the European part. In 1965 the giant Samotlor oil deposit was discovered near the small settlement of Nizhnevartovsk and it became the most important oil production base in the 1980s. The development of Western Siberian oil boosted natural gas production as well. Exports to the satellite states and later to Western Europe rapidly followed. The first long-term contracts with West European countries were signed, marking the beginning of a long-term interdependency. The construction of transcontinental pipelines prompted technological cooperation with Western Europe. In the context of the oil shocks of the 1970s, the countries of the European Community viewed such cooperation positively despite the ideological and political differences. Meanwhile, the East–West trade in raw materials in return for grain and technology marked the end of the autarkic policies pursued earlier by the USSR.

The oil-bearing province of Western Siberia is characterised by a cold, hostile environment with permafrost and peat bogs. The major difficulty for the Soviet oilmen was to build a basic infrastructure in this deserted land and to provide transportation that would bring hydrocarbons over thousands of kilometres to the main consumption areas. To illustrate this, Nizhnevartovsk is situated about 3,000 kilometres north-east of Moscow with winter temperatures that drop to −50°C. The shortage of skilled labour in Siberia was and remains another serious problem. Hence, the success of this project has been regarded with much scepticism both in the West and in the USSR. Despite these difficulties, Western

Siberia became the new oil and gas centre of the country in a very short period of time: production increased ten times from 30 Mt to 300 Mt between 1970 and 1980, reaching a record peak in 1988.

However, the spectacular development of the Soviet oil and gas industry required colossal capital investments, which along with military expenses represented a considerable burden on the stagnating Soviet economy.[10] In fact, the Soviet Union became more and more dependent on the hard-currency earnings from oil exports and subsequently on the world price of crude oil. The oil price counter-shock in 1986 resulted in a substantial loss of hard-currency revenue and aggravated further the socio-economic state of the country already marked by a lack of consumer goods and chronic food shortages. The same year, the Chernobyl nuclear accident had a far-reaching impact on the development of the energy sector as a whole: the construction of nuclear power stations was subsequently cancelled and power generation became even more dependent on hydrocarbons, mostly on natural gas.[11]

Modern Russia had thus inherited a sturdy energy sector developed with massive investments and the work of previous generations of the Soviet people. Recalling Lenin's famous adage, 'Communism is Soviet power plus the electrification of the entire country', it is interesting to note that while communism had turned out to be a failure, the Soviet electricity generation system was working well and quite reliably. The USSR had created adequate power machine building, electrical equipment industries, and advanced science and research. On the other hand, the Soviet industries (mostly heavy industry) accounted for more than half of the total primary energy consumption. If the oil shocks had stimulated the energy conservation programmes in the West, the Soviet industries had little financial incentives to implement saving measures and were characterised by a high degree of energy losses and wastage. Industrial regions also suffered from a variety of environmental problems and pollution.

The 1950s and the 1970s were not only periods of rapid industrial growth that fostered the accelerated rate of fossil fuels production but they were also the golden age of scientific research and innovation in energy (see Box 3.1). However, while the rapid development of the Siberian oil and gas fields provided an abundant supply of cheap fuel, alternative energy research was not considered a priority. After the disintegration of the USSR, the R&D activities of many promising technologies were practically abandoned due to the lack of financial support.

Box 3.1 Energy research and technology

Many technological advances and inventions were a by-product of the active military research pursued by the great powers in the aftermath of the Second World War. The world's first nuclear power station was started up in the town of Obninsk, near Moscow, in 1954. At the same time, Russian physicists Igor Tamm and Andrei Sakharov came up with the idea of a controlled thermonuclear fusion reactor while working on the hydrogen bomb. The first experimental devices for fusion research, baptised Tokamak, were built in the Soviet Union in the late 1950s–1960s. These are doughnut-shaped chambers surrounded by powerful electromagnets where gas is heated to hundreds of millions of degrees Celsius to become plasma so that nuclear fusion can be achieved. While this technology is still not mature, the principles laid out by these pioneering works may provide inexhaustible sources of energy for future generations.

The 1960s were a time of great expectations. The Soviet economy had a long way to go to effectively 'catch up and overtake America',[12] but Russian scientists were narrowing the gap and even taking the lead. Several important breakthroughs in renewable energy and energy efficiency were made at that time and physicists enjoyed enormous prestige in Soviet society. Along with the Americans, the first advanced magnetohydrodynamic generator (MDH) was completed by 1964. Soviet physicist Khristianovitch led pioneering research in the combined gas–steam turbine that became a favourite of the European electric utilities later in the 1990s. In the early 1960s, the geologist Andrei Trofimuk discovered deposits of gas hydrates in the permafrost soil of Siberia. Gas hydrate is an ice-like crystalline substance composed of water and natural gas. Trofimuk was the first to provide a global estimate of the world's reserves of hydrates-bound natural gas found both in the soil and in the ocean bed, which appeared to be measured in astronomical figures (10^{17}–10^{18} bcm). While gas hydrates are too expensive to process today they could provide a vast source of energy in the future.

3 From post-Soviet Russia to modern Russia

In 1991 the USSR ceased to exist. The new government of the now independent Russian Federation opted for radical market reforms based on price liberalisation, privatisation of state enterprises, severe budget constraints and free external trade. The results were immediate: hyperinflation wiped out savings, insiders or criminal structures took control of a large part of the economy and several million workers were not paid for months. Finally industrial production collapsed. In 1992 alone, the GDP fell about 15 per cent: Russia was hit by the worst economic depression probably ever in the industrialised world. Between 1989 and 1998, the GDP shrank by half, electricity generation dropped 30 per cent, natural gas consumption fell about 13 per cent and oil and coal consumption dropped by half. In short, Russia had moved from a state of crisis to one of catastrophe.[13]

The disintegration of the USSR fractured a largely unified energy system, built irrespective of regional borders. Thousands of kilometres of pipelines, major oil terminals and a number of refineries were henceforth located in different independent states. The former Soviet republics, previously interdependent, experienced severe electricity, oil and gas shortages. Exports from Russia were sporadically cut off for chronic non-payment. Ukraine, then a major consumer of Siberian gas, was particularly hit by the energy crisis. In Russia, the whole economy was trapped in a debt spiral: energy firms were owed the most, as large quantities of electricity, oil and gas deliveries had not been paid for.[14] Shortages in jet fuel and gasoline caused transport dysfunctions. In some regions, notably in the maritime province (Vladivostok), winter electricity cut-offs of indebted public services and households became frequent and had dramatic consequences. Grain crops could not be harvested, as impoverished farmers could not afford fuel.[15] The domestic energy supply became a significant issue of national security and a matter of survival for the nation. All this forced the government to adopt the law on the state regulation of tariffs for electricity and heat in 1995. The total debts to national energy providers, however, continued to rise to new record levels. Meanwhile, energy companies accumulated large tax debts to the federal budget and failed to pay salaries to their own workers. In 1996 nearly half a million Russian coal miners went on strike demanding over $200 million in back wages. By the end of 1996, the total debts owed to energy suppliers reached $58 billion. The same year, several well-connected bankers bankrolled Mr Yeltsin's re-election campaign in exchange for the country's most valuable assets, notably in the oil industry. The odd kleptocratic economic system that was neither a free market democracy nor the state-regulated one had been formed with collusion between the Kremlin and the oligarchs.

Paradoxically, it was the financial crash of 1998 that marked a turning point in the development of the Russian economy. The devaluation of the rouble boosted domestic production; the growing world price of oil helped to reduce the external debt and provided the state budget with taxes from export revenue. Under the presidency of Vladimir Putin (2000–8), the 'commanding heights' of the economy were redefined: political power was consolidated and centralised, the state tightened control over strategic sectors while the further privatisation of the energy sector (notably, electricity generation) was pursued along with some other neo-liberal policies. Since 1998 the macroeconomic indicators have improved significantly and political stability has been achieved.[16] However, the radical transformation of Russia had a deep impact on industry and indeed on all spheres of the society.

3.1 The oil sector

Radical changes took place in the oil industry in the years following the dissolution of the Soviet Union. At the initial stage of the reforms, the vertically integrated oil companies were created out of production units and refineries previously under the control of the Soviet Ministry of Oil. That is how LUKOil, Surgutneftegaz, Yukos, TNK (Tyumen Oil Co.), Sibneft (Siberian Oil) and Rosneft (Russian Oil) appeared between 1992 and 1995. During the 1990s, the share of state ownership gradually decreased from 100 per cent to zero in most of these firms. Rosneft remained the only national company fully owned by the state, and was partially privatised only in 2006 when it conducted a successful $10.6 billion initial public offering (IPO). A number of other companies, such as Tatneft and Bashneft were controlled by the autonomous republics of Tatarstan and Bashkortostan (Volga–Urals region). During the loans-for-shares schemes between 1995 and 1996, a small group of insiders and bankers took control of Yukos, Sibneft and TNK for only a fraction of their real market price.[17]

However, the privatisation[18] and liberalisation of the Russian oil sector did not bring about the needed investments. More than that, they failed to modernise the oil sector sufficiently to ensure sustained growth. The overall efficiency of the industry dropped dramatically: while production fell from its peak of 568 Mt in 1988 to 300 Mt in 1998, the number of employees increased from about 130,000 to almost 300,000 workers.[19] In other words, the productivity measured by output per person decreased several times in the last decade. The situation in geology and oil prospecting became deplorable: for example, by 1998 deep exploratory drilling had decreased five fold.[20] At the same time, oil companies (Yukos, TNK, Sibneft) managed by financiers were actively engaged in profit maximisation through ingenious financial schemes and in skimming off the easiest oil that yielded high rates of production increases. These activities coupled with aggressive tax evasion were clearly oriented to boost the market capitalisation for later sell-offs instead of securing the long-term development of the industry. The world crude prices that hit rock bottom in 1998 provided favourable market conditions a few years later. In fact, in 2003 half of TNK's assets were sold to BP in a deal worth $6.75 billion; in 2005 Sibneft was sold to Gazprom for $13 billion. In 2004 the arrest of the CEO of Yukos prevented the mega sale of the strategic stakes of the company to Exxon Mobil for an estimated $25 billion. In the words of Marshall Goldman, the author of *The Piratization of Russia: Russian Reform Goes Awry*: 'if most American robber barons had at least created something out of nothing, the Russian oligarchs added nothing to what already was something'.[21]

By 2002, Russia became a top oil producer again. Nevertheless, the current oil production in Russia is unlikely to reach the peak accomplished in Soviet times and the rhythm of production is slowing down. For some experts, the use of recovery techniques will just accelerate the depletion, as Russia's proven oil reserves are limited (Dienes, 2004). It has also been noted that the recent upsurge in oil extraction mostly comes from oil that was not extracted during the chaotic 1990s. By and large, it is unclear whether Russia will be able to maintain its oil production growth over the next decades given that virtually no new significant oil field has been discovered and that investment in oil prospecting remains abnormally low.

A few years after the second wave of privatisation, the organisation of the petroleum industry again underwent important changes. Mergers and takeovers reconfigured the Russian oil scene, with the following major trends and events:

- The return of the state that progressively gained control of about 40 per cent of oil production.
- State-controlled Rosneft emerged as the top Russian major (2.2 million b/d in 2007) after the acquisition of Yukos's main production unit Yuganskneftegaz in 2005.
- The acquisition of Sibneft by Gazprom in 2005 heralded the transformation of the natural gas monopoly into a global energy player.
- The formation of strategic alliances with foreign oil majors (LUKOil-ConocoPhillips, TNK-BP).
- LUKOil and Rosneft entered the list of the world's top ten companies in terms of proven reserves;[22] Gazprom, Rosneft, LUKOil and Surgutneftegaz also became the largest companies in terms of market capitalisation with overall value reaching almost half a trillion

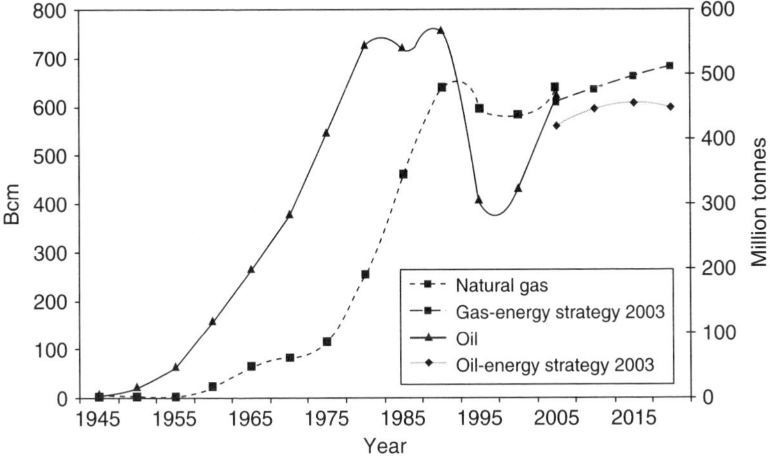

Figure 3.2 Oil and gas production in Russia

US dollars in 2007. This figure, however, shrank dramatically with the stock market crash two years later.[23]

3.2 The gas sector: from the Soviet gas ministry to the Gazprom empire

Omnipresent Gazprom currently dominates the Russian energy scene: it accounts for 85 per cent (as of 2006) of gas production and has expanded its activities both domestically and abroad. Gazprom is increasingly extending its presence in the EU gas market, through joint ventures and equity positions in every part of the gas value chain. The European Union as a whole gets about a quarter of its gas from Gazprom. Besides the traditional gas business, the company has been actively acquiring assets in oil, petrochemicals, electricity generation and coal. The company's objective is to 'become one of the largest integrated energy companies in the world, spanning oil, gas and electricity'.[24]

The evolution of Gazprom over the last fifteen years has been remarkable. Gazprom was created from the Soviet Ministry of the Gas Industry and has become one of the top energy utilities firms in the world. In contrast to the oil industry, it has not been divided into several companies but kept as a single unit. The company provides about 20 per cent of earnings to the federal budget and it enjoys a particularly intimate relationship with the political power. Mr Tchernomyrdine, the Chief Executive of Gazprom, became Prime Minister of Russia (1992–8). Mr Medvedev, a protégé of former President Putin who served as the chairman of Gazprom's board of directors, was elected Russian president in 2008.

During the depression period, the non-payments for gas deliveries severely undermined the company's finances and had a negative impact on its investment policy. Low domestic prices of gas were kept heavily regulated and Gazprom literally subsidised Russian customers throughout the last decade. It is natural gas that is considered to have kept the Russian economy afloat during the economic crisis. Moreover, until recently Gazprom provided the former Soviet republics with gas at prices considerably lower than those paid by EU importing countries. Accumulated debts and accusations of stealing the gas destined for the EU market became a source of growing tensions between Russia and its neighbours (notably, Ukraine).

At the end of the 1990s, Gazprom was submerged by a number of corruption scandals that seriously tarnished its reputation. By the beginning of the 2000s, sweeping changes in the top management brought the company under the control of the newly elected president Vladimir Putin. In 2004, the state increased its stake in Gazprom from 38 per cent to a controlling 50 per cent plus one share. At the same time, the restrictions on foreign investments were lifted and the company became fully open to foreign investors. Private shareholders and companies such as E.ON Rurhgas now own the other half of its shares. The federal law approved by the state Duma also granted Gazprom exclusive rights to export natural gas.[25]

The Russian gas scene could evolve in the future with the dynamic production growth coming from the so-called 'independent gas producers' such as Novatek and Itera, or oil majors seeking to develop the gas business (Stern, 2005). In any case, Gazprom is targeted to remain a major actor of the Russian energy scene, while fully asserting itself as a global energy player.

3.3 The electricity sector: restructuring the state monopoly

Since Lenin's nationwide electrification plan (GOELRO), power generation has been considered fundamental for the economic development of the country. The Soviet power generation system was highly centralised and brought about the development of the regional joint power systems with the objective of accomplishing the Nationwide Unified Power System. The vast size of the country and the remoteness of the consumption centre from the major fossil fuel deposits made it necessary to develop efficient long-distance electric power transmission technologies. This also enabled the transfer of energy during peak loads between different time zones. The Soviet Union became a world leader in high-voltage transmission and a pioneer in Ultra High Tension (UHT) engineering. For example, a UHT (1200 kV AC) line connected the powerful coal-fired stations in Kazakhstan to the European part of Russia.

In the 1980s, the USSR introduced new capacities at a pace of 5–10 GW per year. Substantial reserve capacities guaranteed a high level of security and reliability of the grid. The collapse of the USSR caused a clear rupture in the development of power generation: electricity production in Russia dropped from 1082 TWh in 1990 to 827 TWh in 1998. With the rebound of the economy since then, electricity consumption is expected to achieve the 1990 level in 2008. However, the current installed capacity of 220 GW (in 2006) has practically not increased in the last fifteen years and the bulk of the power generation sector is operating on outdated and worn-out equipment. For instance, the failed transformers that caused the most serious blackout in Moscow in 2005 were over forty years old.

Russia generates 43 per cent of its electricity from natural gas, followed by hydropower (18 per cent) and nuclear (16 per cent). The accelerated development of gas production in Western Siberia and the construction of transcontinental trunk lines to the West resulted in a significant shift to natural gas in the electricity generation mix during the 1980s. This had a positive environmental impact in the European part of the country. The share of coal and oil decreased gradually. The share of coal-based electricity generation is high only in Siberia, where major Russian coal producing regions are located.

Until its restructuring, the key player on the electricity market was the Unified Energy Systems of Russia (RAO UES). The state-controlled monopoly RAO UES was established in 1992. It provided about three-quarters of Russia's total electric power output and owned most of Russia's transmission lines. Since 1998, the holding was managed by Anatoly Chubais: well respected in international financial circles, he is certainly the most controversial public figure in Russia as he was responsible for the ill-fated privatisation programme of the early 1990s. 2001 marked the beginning of the major restructuring of the Russian electricity industry, which was completed in 2008. Based on the British model, the large-scale liberal reform implies the full corporate separation of monopoly network activities (long-distance high-voltage transmission and local grids) from power generation and retail supply. The state retains control of hydro plants and the nuclear power industry. The network business remains a regulated monopoly, while the generation, sales and repair activities become competitive. The generation utilities of the monopoly have been repackaged into twenty-one generating companies (fourteen 'Territorial

Generation Companies' (TGK), six 'Wholesale Generation Companies' (OGK) and one Hydro OGK). Most of them were privatised with a series of floats on the stock exchange in 2007–8. Most of the control stakes were bought by Gazprom and other large Russian industrial groups. Some European utilities (Fortum, E.ON, Enel) also took ownership of the generating companies. The new owners have committed to a large-scale investment programme: 29 GW of new generating capacities according to Chubais' investment plan between 2006 and 2010. Given that very little new generating capacity has been commissioned so far (at a rate of less than 2 GW per year since 2000), this target seems unrealistic. Moreover, the deteriorating state of the economy since mid to late 2008 will certainly impact the development of the power sector.

It is argued that the sell-off of the monopoly is a necessary step to attract capital that is sorely needed for the modernisation of the power industry. In contrast, opponents of the reform believe that the break-up of the unified power system is just a round-about way of redistributing the assets to enrich the happy few. Given the notorious experience of market deregulation in the past, there are risks that such restructuring may result in a speculative outburst on the electricity market with more blackouts and price hikes in the future.

3.4 The Energy Strategy and Russian-style 'resource nationalism'

In 2003, the Russian parliament adopted a new Energy Strategy, drafted by the government and experts from the industry and academia (Russian Government, 2003). Taking into account the new context of economic growth and the rising oil prices, this legal document set out the main guidelines for the development of the sector until 2020. Two alternative scenarios for different rates of economic development and world crude oil prices were provided with forecasts for production, consumption and exports of raw materials and electricity for each branch of the industry. The fuel and energy complex is regarded as the basis of economic development and the instrument for both internal and external policies. The formation of a 'civilised energy market' has been recognised as essential to accomplish the competitive energy sector with the state serving as a regulator: 'Energy strategy is firstly an ideology, and secondly – numbers.'[26]

The strategy underlines the central role the state should play to ensure the most effective exploitation of Russia's mineral wealth, through price regulation, taxation and investment policies, and the legal framework. The state should promote national companies that serve the national interests first: 'Ongoing global competition to gain control over hydrocarbon reserves has shown that state-owned and backed companies have considerable advantages in obtaining dominating positions on international markets.'[27] One striking feature of the Russian fuel and energy sector at the beginning of the twenty-first century is undoubtedly the emergence of oil and gas operators with an international dimension in which the state has a majority control (Gazprom, Rosneft).

In addition to the growing presence of the state in some of the leading energy companies, the government has imposed stricter conditions for accessing the country's resources: the Subsoil legislation was modified several times to grant the federal government the sole authority to issue mining permits. Moreover, the adoption of a list of 'strategic fields' in 2007 now grants state control over important oil and gas fields, such as the Chtokman in the Barents Sea. Gazprom and Rosneft, two Kremlin-controlled companies, are often privileged in winning the licence permits.

In the West, the recent developments such as in Sakhalin[28] are generally interpreted as new cases of 'resource nationalism'. This term designates in fact the restricted access for multinationals in developing countries to natural resources. It is generally accompanied by the growing weight and influence of national, often state-controlled companies that control the majority of oil and gas reserves.[29] However, the use of such a term in the Russian context requires closer examination. Historically, the fuel industry has always been considered as a key national economic sector that enjoys special attention from the government. In this sense, the anarchic period of Yeltsin appears to be an aberration. It is worth recalling that during the Yeltsin presidency state institutions became weak and corrupted while the real power was concentrated in the hands of the president.[30]

It is interesting to learn what Russians think about it: according to polls conducted in 2005–8 by VTsIOM, the leading public opinion research centre, 51 per cent of respondents would like the results of the privatisation to be revised, 56 per cent support the further reinforcement of the state, 45 per cent agree to the idea of nationalising the entire oil and gas industry and 66 per cent believe that foreign capital should not be invested in the energy sector.

4 The investment and technology challenge

4.1 Exploration in the oil and gas sectors

The crucial questions for Russia today are to know how long it can sustain current production levels from mature fields in decline and where the new sources will come from in the future. The new extractive regions are located further north (Yamal peninsula, Arctic Ocean) and further east (Eastern Siberia). These areas are even more hostile and remote than Western Siberia, where the cost of extraction and transportation is already considerably higher than in most other producing countries. It is essential to note that overall geographic and geological conditions for the extraction of hydrocarbons in Russia are worsening: future production will increasingly come from harder-to-reach reserves. Thus, the development costs will increase significantly in the next two decades.

In the gas sector, over two-thirds of the production come from three giant fields (Yamburg, Urengoï and Medvezhie) and have entered into a declining phase. One of the greatest challenges is to develop fields in new gas provinces of the Yamal peninsula and the Arctic shelf. The Chtokman field, for instance, is one of the world's largest natural gas deposits. It is particularly challenging: not only is it 300 metres deep but the nearest seaport of Murmansk is 600 km away. Icebergs are frequent in these northern latitudes and pose a serious threat to offshore platforms. Certainly, this project will be the world's most challenging and costly development. Its successful realisation is not possible without broad international cooperation.

In the petroleum sector, the oil production in Western Siberia has reached a plateau. The upsurge in oil production between 2000 and 2006 originated from deposits exploited at the end of the Soviet era or from idle fields. It is clear that such growth is not sustainable. To maintain current extraction levels, new capacity and new fields must be developed in the very near future. According to the Russian Energy Strategy, East Siberia and the Far East will largely contribute to the increase in output in the next decade. Yet, the development of new fields in those regions will require a radical expansion in prospecting activities and in the development of a transport infrastructure in the Siberian desert. Moreover, these new fields are generally smaller and more dispersed, thus requiring colossal investments. 'To explore and to develop the rich deposits of Siberia are as difficult as to explore the space. Here everything is unique and tremendous,' stated Mikhail Lavrentiev, founder of the Siberian branch of the Russian Academy of Sciences in the 1960s.

Great uncertainty surrounds the estimation of oil reserves. For example, the *Oil and Gas Journal* 2007 estimates the proven reserves at 60 billion barrels (8.2 billion tons), while the International Energy Agency (2002) forecasts 146 billion barrels (19.9 billion tons). In fact, only a small fraction of the huge Russian territory, spanning over 17 million square metres, has been explored. According to Andrei Trofimuk, only a quarter of the country's subsoil has been explored. Little is known about Eastern Siberia, or the Arctic shelf, where exploration has just started. Exploration activities were almost abandoned and drilling considerably reduced during the 1990s, after the state ceased to finance them in 1992. Moreover, oil reserves are now considered as strategic assets and have thus been classified as a 'state secret' since 2003.[31]

The Russian Ministry of Energy has estimated that $90 billion of investments will be necessary for exploration by 2015. Yet, the amount invested by oil and gas companies remains far below this figure. Flexible tax policies that would encourage the development of exploration and development of difficult fields should provide incentives for companies. The stakes are high as Eastern Siberia may have enormous hidden wealth in its depths.

4.2 Transport networks and environmental concerns

Russia is criss-crossed by a web of hundreds of thousands of kilometres of pipelines, most of them built in Soviet times. Obsolete equipment and human errors cause a growing number of technical failures and accidents. In northern regions, the ecosystems are more fragile and coexist uneasily with polluting industries such as fossil fuel extraction. Scenes of spilled oil along the trunklines, burning torches of oil and gas wells and traces of trucks stuck in permafrost that will remain for decades are sad reminders of the activities of the oilmen. Wear-and-tear due to corrosion leads to frequent ruptures that cause not only heavy environmental damage but also tragic accidents. In 1989, an explosion from a leaky liquefied gas pipeline caused a train disaster that killed more than 500 and wounded over 600 passengers.[32] In 1994, a leak from a ruptured pipeline caused a large-scale environmental disaster on the territory of the Komi Republic (north-west of Russia). The total volume of oil spilled is reported to be between 14,000 and 100,000 tons. The situation has hardly improved in recent years: about 10,000 tons of crude oil is lost annually due to thousands of accidents from ageing pipes. As oil production is increasing, so are the numbers of oil spills. According to the official statistics recorded between 2000 and 2004 there were 10,647 pipeline accidents with more than 27,000 tons of oil spills in the Khanty-Mansy region alone in north-west Siberia. The number of reported accidents has almost doubled since 2000.[33]

The bulk of oil exports (roughly 1.3 million b/d in 2007) are transported by the Druzhba ('Friendship' in Russian) pipeline. This is the longest trunk pipeline in the world and it was built in the 1960s to supply crude oil to the 'fraternal' states of the Eastern bloc. Today it needs significant modernisation. Any disruption caused by an accident on this pipeline could wreak havoc on oil markets. High volumes of oil have also been shipped through Black Sea ports to the Mediterranean through the narrow Bosphorus Straits. The risk of oil-spills in this vulnerable and densely populated area is high and any major tanker accident could have disastrous consequences. Transneft, the national oil pipeline operator that owns most of the country's pipeline system, is involved in several multibillion dollar projects aimed at developing new export routes and upgrading the existing pipelines. The planned joint Druzhba–Adria and Burgas–Alexondropolis projects will bypass the Bosphorus Straits. Part of the oil transported by Druzhba is being redirected to the Baltic Pipeline System.

In the gas sector, over half of the transmission network is more than twenty years old. The infrastructure requirements entail upgrading and expanding the export pipeline capacity to Europe. Consequently Gazprom has been making significant investments to modernise existing infrastructures and to build new transport routes.

4.3 Private–public and foreign investments

According to the Russian Energy Ministry, the total investments required in the energy sector should amount to around $660 and $810 billion from 2002 to 2020. The International Energy Agency estimated that over $1 trillion is necessary. Such figures require the mobilisation of both public and private funds, including foreign investments. Undoubtedly, the vital challenge for the Russian energy sector is to define new investment policies. Without a long-term investment strategy Russia's future as an energy power and its capacity to honour its long-term contracts with its European customers will be challenged. In this sense, the energy future of Russia will depend on the outcome of public–private cooperation as well as cooperation with foreign investors.

During the 1990s, foreign investment remained extremely low due to the weak protection of property rights, the baffling tax legislation and rampant corruption. Meanwhile, capital flight amounted to $150–$300 billion between 1992 and 1999.[34] Some of this capital started returning to Russia later on, making offshore havens such as Cyprus for top foreign investors. As the economy rebounded and the legislation improved, foreign investment gradually started growing, amounting to $121 billion in 2007. Still, the share of foreign direct investment remained low (about $50 billion in 2007) with the bulk of it going to the oil and gas sectors.

Investments in developing new fields often in partnership with foreign majors have progressed as well. In fact, the dwindling reserve base and soaring prices have stimulated multinational oil corporations to look for profits in regions where costs and risks are higher. Oil majors started facing fierce competition from national companies backed up by their respective governments. In this context, Russia seems to be more open than most other producing countries such as Saudi Arabia, Venezuela or the United States.[35] This is why BP, Total and Shell are striving to gain access to Russian fields, despite unfortunate past experiences and tighter rules.

National companies have also increased their interest in domestic investment. In 2005, Gazprom adopted an $11 billion yearly investment programme. However, the bulk of this sum was mainly directed at foreign acquisitions, pipeline projects such as Nord or South Stream or investments in power generation. The company also spent $13 billion on the acquisition of the oil company Sibneft, making its ex-owner a multibillionaire. Even if Russian oil and gas companies today possess substantial financial resources thanks to high commodity prices, the way they are utilised often seems inefficient and mysterious.

5 Climate change policies and the Kyoto Protocol

During the crisis in the 1990s Russia's CO_2 emissions were considerably reduced: for example, the emissions from fossil fuel decreased by one-third. Still, Russia has a fairly high rate of carbon dioxide emissions (about 12 tons per capita) and it accounts for 17 per cent of the world's carbon dioxide emissions. In 2004 the state Duma ratified the Kyoto Protocol, and thus has committed to stabilising its emissions at the 1990 level. Given the contraction of greenhouse gas emissions during the 1990s, Russia may benefit in the short term from a surplus estimated at between 330 and 800 Mte CO_2.[36] The Kyoto mechanisms could provide an additional stimulus to enhance national energy efficiency. Indeed, a potential exists for energy savings. For example, in the natural gas sector, methane leaking from the transmission distribution systems accounted for 11.5 bcm in 2004. Upgrading the transmission system through more efficient compressors could save billions of cubic metres of gas alone. In the oil industry, the flaring of the associated gas amounts to at least 15 bcm per year. Yet, some Russian economists and climate experts have harshly criticised this decision on the grounds that it is politically motivated and would threaten economic growth. Yuri Izrael, the vice-chairman of the UN Intergovernmental Panel on Climate Change (IPCC), wrote: 'The Kyoto Protocol is economically hazardous to Russia ... the Kyoto Protocol is scientifically ungrounded and does not indicate the road towards the end set. The economically inefficient Protocol will lead to only an insignificant cutting of the hothouse emissions.'

6 Russia as a major energy consumer

As an industrialised country with a population of 144 million, Russia belongs to the club of the few privileged major energy-consuming countries. Its consumption pattern is largely defined by its northern climate, vast territory and the structure and quality of its industries. Indeed, vast and sparsely populated countries such as Canada or Australia tend to have elevated energy consumption per capita. Northern counties such as Norway or Sweden are also characterised by a high rate of primary energy consumption per inhabitant that provides comfort and high living standards for their populations. The situation in Russia is somewhat different: while the average primary energy consumption per head is comparable to that of other industrialised nations, the welfare of its average inhabitant, measured by GDP per capita, is lagging behind. Russia has notoriously high-energy intensity industries, partly because energy-intensive industries such as the metallurgy and the chemical sectors still represent a significant share of the total industrial output. The deplorable state of outdated equipment is only part of the problem. For example, the average Russian metallurgical plant needs at least twice more energy to produce one ton of steel than its German counterpart. Former Prime Minister Viktor Zubkov[37] acknowledged that: 'There [in energy savings] we have a large, unfortunately, unused potential. We are indecently wasteful and energy saving capacity is estimated to amount to 45 per cent of our total energy consumption. One-third of our fuel and energy resources is being lost or used inefficiently.'

As the energy demand increases with the strong economic growth since 1999, the capacity of the energy sector to meet the growing needs is questionable. In particular, worries of an impending gas shortage have surfaced, due to the higher than expected growth in electricity demand.[38] A supply gap may occur due to the rapid decline of fields in production and the delay required to put new fields into operation: for instance, the gas production from the Yamal peninsula will not begin before 2011, and that of Chtokman in 2012 at the earliest. This is why Gazprom counts on gas imports from Central Asia (see section below).

In fact, the power sector drives demand with the increase of gas use and gas-fired plants. These have a relatively low up-front investment cost compared to nuclear generating capacities, which have high initial investment costs. The share of gas in Russia's primary energy consumption has expanded from 42 per cent to 54 per cent since 1990, while oil and coal decreased their contribution to 19 per cent and 16 per cent

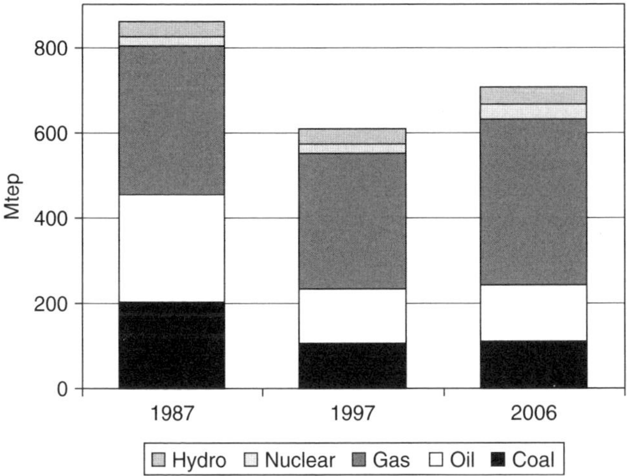

Figure 3.3 Russian primary energy consumption by fuel
Source: BP Statistical Review 2007.

respectively. Today about half of the electricity comes from gas-fired power stations (Figure 3.3).

If Russia could reduce its dependence on gas, and develop coal and nuclear sources it would help to re-equilibrate the primary energy consumption and free more gas for exports. The federal agency for nuclear energy (Rosatom) plans to build 40 nuclear reactors over the next 25 years, which would bring the share of nuclear up to 25 per cent (against 16 per cent in 2007). However, the nuclear programme appears to be very ambitious and it is questionable whether Russia has the necessary means and the qualified manpower to reach such targets (see Box 3.2).

In 2006 the Russian government made the decision to implement the gas market reform after a decade of cheap gas: prices for gas are being gradually increased to reach the European levels (transport and taxes deducted). The high prices should stimulate the adoption of energy-saving technologies in industry and should provide incentives for companies to develop new gas fields. In particular, it will make the development of the Yamal peninsula by Gazprom economically meaningful.

Yet higher prices may not necessarily have an impact on the demand side, as the price elasticity remains low. Take the example of district heating. Higher gas bills will not necessarily spur energy savings for households. First, most of the heat losses occur in the old district heating networks. Second, in most of the apartments it is not possible to regulate the room temperature so it is not uncommon for the room temperature to be +25°C while it is −25°C outdoors! In fact, the whole infrastructure needs to be updated and renewed, in both the industrial and residential sectors. This is tantamount to energy saving. Besides, raising household energy prices could have dramatic social consequences for the low-income population (one-fifth of the Russian population still lives well below the poverty line). The price policy reforms could have another far-reaching impact for the EU: as the domestic market becomes more profitable, the incentives for exports to Europe will decrease. Overall, the implementation of energy conservation and energy-efficiency policies remains crucial for the country's future. Russia's energy sector is not sustainable without massive investments to improve the country's energy efficiency. So far, private and public investments in this area are insignificant and governmental regulations are virtually non-existent.

Box 3.2 Nuclear energy in Russia

Russia has a very ambitious plan for expansion of its nuclear industry. Today, there are about 23 GWe of nuclear generating capacity in operation in Russia. The national plan approved in early 2008 calls for about 52 GWe in 2020[a] and a prominent national laboratory recently proposed scenarios of roughly 230 GWe in 2050 and 540 GWe in 2095, including a large number of sodium-cooled fast breeder reactors beginning in 2025.[b] In addition, Russian industry hopes to export as many reactors as it will build at home and offers to take care of reprocessing the spent fuel.

However, there are some doubts about the implementation of this plan, at least within the time frame envisaged. Russia inherited a large nuclear industry, both civilian and military, with little distinction between the two. It includes high-level research institutes, factories, some fuel reprocessing facilities, excellent and large capacity for uranium enrichment, and a little over 20 GWe of operating nuclear power plant capacity. But the dynamics of the Soviet Union's nuclear industry eroded after the Chernobyl accident in 1986 and the dissolution of the USSR in 1991. There was no new construction of nuclear plants, some of the construction under way was stopped, and in practical terms there was no new hiring in the nuclear industry at large.

In 2005, the Russian government, under President Vladimir Putin, decided to revitalise the nuclear industry. It created the Federal

Atomic Energy Agency, known as Rosatom, under the direction of Sergey Kirienko, a former prime minister, which included both military and civilian activities. In 2007, Rosatom became a corporation. The government also created a new joint stock company called Atomenergoprom to act as a holding company for all Russian nuclear enterprises, including the nuclear power producer, Energoatom. AEP will remain majority-owned by Rosatom (which currently owns 100 per cent of its stock) but will be able to finance its activity through the market and outside investors.

Some sceptics[c] believe that the goal of expanding nuclear power will not be achieved in Russia due to lack of young engineers, scientists and skilled workers involved in nuclear energy, since other sectors pay better and are more prestigious. In addition, they cite the relative obsolescence of many manufacturing facilities involved in the nuclear industry. They also note that under the new management team, the proposed nuclear power production goals were not achieved: five unfinished nuclear power units were supposed to be completed and put into operation by 2007; instead, there were only two. In addition, the capacity factors of existing plants were not improved as much as had been planned.

Nevertheless the Russian nuclear industry still has some advantages that could facilitate future development. Russia has a 1200-MW pressurised water reactor design similar to the first two Russian nuclear power units built in China and now operating. This design is safe and robust and belongs to the category of third-generation reactors alongside the Areva EPR or the Westinghouse AP1000. Russian industry is building two similar reactor units in India and is completing one in Iran, and has a contract to build two in Bulgaria.

There is also very clear interest, from very powerful sectors of the Russian economy, notably Gazprom, to finance and develop nuclear energy. Indeed, Russia has many gas-fired power plants, and replacement of their generation with power from nuclear plants or avoiding the need to build new gas-fired plants would allow Gazprom to export more gas for prices that are already high and are sharply increasing, making investment in nuclear profitable in spite of large increases in capital costs over the past few years.

We may also mention two specific strong points of the Russian nuclear industry. One is the capacity to enrich uranium. The Soviet planners made the right choice of gas centrifuge enrichment technology early on in the 1980s, unlike the US and France which continued with the much less efficient gaseous diffusion technology and are only now building centrifuge enrichment plants. Russian industry currently supplies some 40 per cent of world enrichment services.

Russia's other strong point is sodium-cooled fast breeder reactor technology, which represents the future of nuclear energy within a few decades if nuclear energy use is to expand. The Russians built several experimental and prototype fast breeder reactors and have been operating an industrial prototype of 600 MWe for more than 20 years with a capacity factor that is among the best of all Russian nuclear power plants. They are building another fast breeder reactor, called BN-800 (800 MWe) that is planned to operate in 2012. The design of a 1600–1800 MWe FBR is on the drawing board. The Russians have therefore one of the best, if not the best, experience worldwide in building and operating sodium-cooled FBRs.

In summary, it is too early to judge the effectiveness of the new organisation and management teams. But it is likely that Russia will develop its nuclear industry (nuclear power in Russia and exports, including fuel cycle services), although perhaps more slowly than current plans indicate.

C. Pierre Zaleski (CGEMP).

[a] V. Rachkov, Director, Scientific Policy, Rosatom. Moscow, June 2008.

[b] N. Ponomarev-Stepnoy, Vice President, Kurchatov Institute. Moscow, June 2008.

[c] B. Nigmatullin, First Deputy Director, Institute of Natural Monopolies and former deputy minister of nuclear energy, and M. Kozyrev, editor, Russian edition, *Forbes*. *ProAtom,* May 2008.

7 The Russian export strategy: new routes and new markets

7.1 Exports to Europe and transit issues

For more than forty years, Russia has been steadily stepping up its oil and gas sales to the EU countries. Today, the enlarged EU-27 relies on Russia for a quarter of its gas needs and one-third of its crude oil imports. While the EU has become increasingly dependent on Russian deliveries, Russia has also become very dependent on European markets: the EU currently takes about 70 per cent of gas and 80 per cent of oil exports from Russia.[39] The trade, however, is asymmetrical: Russia imports equipment, consumer goods and high value-added products, while it exports raw materials.

Following the break-up of the Soviet Union, Russia became dependent on its western neighbours, notably Ukraine, Belarus or the Baltic States for oil and gas transit. In the 1990s over 90 per cent of Russian gas exports to the EU transited through Ukraine, which became a key energy transit country. Major Soviet oil terminals such as Ventspils (Lithuania) or Odessa (Ukraine) and refineries are located in the newly independent states. As a grave economic crisis hit all of the ex-USSR republics, the demand for energy fell and the resource-poor transit states were not able to pay for Russian oil and gas. Note that during the first post-Soviet decade, Russia provided hydrocarbons to former Soviet states at a much lower price than that paid by Western or Central European countries. Recurrent non-payment for oil and gas deliveries, the poor state of the transport network, plus high transit fees and disputes over fuel prices prompted Russia to diversify its export routes. Political issues such as the pending entrance to NATO of the Baltic States and the discrimination against the Russian-speaking population have influenced the decision to develop transport hubs on Russian territory as well.

Crude oil, which was traditionally transported through the Baltic States, has been re-routed through the new Baltic Pipeline System (BPS). The BPS is linked with the new giant oil terminal of Primorsk (about 1.5 million b/d in 2007), near St Petersburg, thus giving a direct outlet to the northern European markets by the Baltic Sea. Shipment of oil through the northern seaport of Murmansk, located above the polar circle, has been growing as well. New gas pipeline projects such as Nord Stream (under the Baltic Sea), Blue Stream or South Stream (under the Black Sea) bypass any third transit country. The realisation of these costly and complex infrastructure projects would not be possible without the participation of energy giants such as German E.ON, BASF or Italian ENI. Joint mega-projects such as Nord Stream only reinforce the existing mutual dependency between the EU and Russia. Nevertheless, the traditional European markets will not remain the only destination for Russian exports in the long term.

7.2 The eastern vector of Russian oil and gas exports

Russia is increasingly looking to diversify its exports eastwards, notably towards the booming Asia-Pacific region (APR).[40] The first reason is purely geological: vast untapped reserves are to be found in the east, such as on Sakhalin Island, close to Japan or to South Korea. The giant gas field of Kovykta located in the Irkutsk region is much closer to China than to

Germany. Second, the most dynamic and energy hungry economies such as China are in the east. Moreover, sparsely populated Eastern Siberia and the Far East remain the most economically backward parts of the country and new oil and gas projects would boost their development. Finally, some top Russian oil managers have increasingly criticised the predominance of the EU market. Semyon Vainshtok, the chairman of Transneft, remarked: 'Our entire export potential is geared towards Europe, which is overdone with Russian oil ... Today Russia has a chance to open the Asian-Pacific market.'[41] In fact, the Energy Strategy (2003) forecasts that by 2020 the share of APR markets in Russian oil exports should increase from 3 per cent (2003) to 30 per cent (91–105 Mt) and gas exports should rise to 15 per cent (40–2 Bcm).

The most important transport project now underway is the East Siberia–Pacific Ocean (ESPO) pipeline. It is designed to carry up to 80 million tons of crude oil annually to the Pacific coast, as well as to China. The construction of a gas pipeline through the Altai region to China has been planned as well. Yet, the ambitious plans to diversify export markets raise the question of the industry's ability to increase production capacity in Eastern Siberia fast enough. More generally, it is not clear if Russia can meet its growing domestic demand as well as its export commitments and targets. At the same time, Russia's southern neighbors in the Caspian region are increasing their production at a rapid pace to become important suppliers of oil and gas to Western markets.

8 The new independent states around the Caspian Sea

Kazakhstan, Turkmenistan and Azerbaijan are the newly independent Turkic states around the Caspian Sea. Together they hold about 4 per cent of world proven reserves of natural gas and 4 per cent of oil (less than neighbouring Iran). These new states are awash in oil and gas and have recently emerged from obscurity to find themselves at the centre of new rivalries between major world consumers. Having similar political systems (all three are governed by autocratic rulers), these states are very different in terms of size, population and economy (see Table 3.1). Among them is Turkmenistan, the biggest exporter of natural gas. Its gas production is rapidly increasing but it has not yet reached the peak achieved in Soviet times. Azerbaijan is the world's oldest oil-producing region. Its Baku fields that once provided half of the world's oil output have witnessed a long period of decline. Now Azerbaijan is back on the energy scene thanks to oil from the offshore fields, developed by a consortium of British and American companies. Kazakhstan has the largest

Table 3.1 Main economic indicators for Russia, Kazakhstan, Azerbaijan and Turkmenistan (2006)

	Population (million)	*GDP (billion USD)*	*GNI per capita*	*Proven reserves of oil (billion barrels)*	*Production of oil (million b/d)*	*Reserves of gas (trillion cubic metres)*	*Production of gas (billion cubic metres per year)*
Russia	144	987	5770	79.5	9.8	47.65	612.1
Kazakhstan	15	81	3870	39.8	1.4	3.00	2.9
Azerbaijan	8.5	20	1840	7	0.7	1.35	6.3
Turkmenistan	5	10.5	650	0.5	0.16	2.86	62.2

Sources: World Bank, BP Statistical Review 2007.

proven reserves of hydrocarbons. Its gas production has increased threefold since 1991, and its oil production is forecasted to double over the next decade. Yet, Kazakh oil is not easy to extract. Take the example of Kashagan, a super-giant offshore oil field discovered in 2000. The field is so complex both technically and geologically that the overall production cost, estimated at $136 billion, makes it the world's most expensive oil project. Besides, the crude oil from Kashagan is sour, with a high content of toxic pollutants and the development of the field may have a very negative impact on the environment of the Caspian region.

Locked in the heartland of Eurasia, the newly independent states face a similar dilemma: how and where to export their oil and gas wealth with maximum profits. The United States, China and Russia apparently share similar concerns, and have become increasingly active in courting these republics. The US government vigorously supports pipeline projects that bypass Russia and Iran. It backed the Baku–Tbilisi–Ceyhan (BTC) pipeline that now brings Caspian crude oil to the Mediterranean port of Ceyhan in Turkey. The US equally seeks to develop a strategic partnership with Kazakhstan where American companies participate in several mega-projects. The decision of the Kazakh government to join BTC was indeed warmly welcomed by the United States. Likewise, China is actively seeking to secure direct energy supplies from the region. In 2005, the Kazakhstan–China oil pipeline became operational. Two years later, the Chinese oil corporation CNPC started building a pipeline that would transport Turkmen natural gas to China. Contrary to the 1990s, Russia's current strategy aims at securing and consolidating its position as a transit state for oil and gas from this landlocked region. Recently, Russia seems to be regaining its influence, and has concluded several long-term strategic gas cooperation agreements with Kazakhstan, Turkmenistan and Uzbekistan. In fact, Gazprom counts on imports of large volumes of gas from Central Asia as its own production stagnates. Without it, the Russian giant may not be able to honour its export contracts and meet the growing domestic demand. Turkmen's gas fields are relatively easy to exploit, but the transport remains expensive and its obsolete infrastructures need upgrading. However, the Central Asian governments keep raising the price of gas, which may have an impact on future export projects. Given the ultimate pragmatism of the leaders of those states, it is rather difficult to foresee future alliances between the Caspian producers and the major consuming countries.

Notes

1. A peak production of 11.6 million tonnes (85.2 millions barrels) was achieved in 1901. That represented slightly more than 50 per cent of the world's output.
2. Data for 2005, Eurostat (2007).
3. TEK is the transliteration for the Russian abbreviation of Toplivno-Energetichesky Kompleks.
4. Data from Rosstat (2006).
5. 'Russia in the Shadow, Fifth Article: The Dreamer in the Kremlin', *New York Times*, 5 December 1920.
6. Foreign capital was mostly present in the oil industry, where concessions were granted to foreign companies in the 1920s.
7. Ivan Gubkin (1871–1936) developed a theory on the origins of oil and laid out the principles of oil geology in *The Study of Oil* (1932).
8. According to Daniel Yergin (1991), this resulted in reducing the royalty revenues of governments from the Middle East, which was one of the driving forces behind the formation of the Organization of Petroleum Exporting Countries (OPEC).
9. Andrei Trofimuk (1911–99) contributed to the discoveries of major oil deposits in the Volga–Ural region and in Siberia.
10. In 1988 the energy sector accounted for about 15 per cent of the Soviet budget. The share of capital investment in the oil industry increased from 30 per cent in 1970 to 50 per cent in 1988.
11. In the 1980s, the USSR was planning to double its nuclear capacity by the mid-1990s.
12. The motto coined by Nikita Khrushchev, First Secretary of the Communist Party of the USSR (1958–64).
13. This expression is attributed to economist and academician Nikolai Petrakov.
14. The World Bank estimates that Russian fuel- and energy-producing companies were owed about 44 trillion roubles ($9.4 billion) as of January 1996. Only 77 per cent of all the energy delivered to Russian consumers in 1995 has been paid for.

15. As reported by the World Bank, the grain harvest was 22 per cent down in 1995 compared to 1994. Fodder crops also dropped by 36 per cent. The poor harvest was due partly to the fact that farms were unable to buy adequate supplies of fuel and fertiliser.
16. The GDP increased 43 per cent between 2000 and 2007 and energy consumption rose 10 per cent.
17. The government auctioned off substantial packages of stock shares in some of its most important enterprises, such as energy, telecommunications and metallurgical firms, for bank loans. For example, Mr Khodorkovski acquired the shares of Yukos for about $160 million in the auctions held in November–December 1996. As of August 1997, the market price was estimated at over $6 billion. By 2004, the market capitalisation reached $24 billion. See Khlebnikov (2001).
18. The Russians quickly transformed the term 'privatisation' (*privatizatsia*) into 'grab-it-isation' (*prikhvatizatsiya*).
19. Glaziev et al. (2003: 150).
20. Ibid.
21. Marshall I. Goldman, 'Putin and Oligarchs', *Foreign Affairs*, November–December, 2004
22. *Oil & Gas Journal*, 17 September 2007.
23. *Financial Times*, Global 500, 30 March 2007.
24. Alexeï Miller, *Financial Times*, 12 July 2005.
25. The Federal Law on the Export of Gas, July 2006.
26. Victor Khristenko, Minister of Industry and Energy, 4th All-Russian Energy Forum, 3 April 2006.
27. Gazprom meeting, June 2006.
28. In 2006, Gazprom took a majority control in Sakhalin II for a $7.45 billion deal while foreign companies agreed to halve their stakes. See Campaner (2007b).
29. It is estimated that national companies control 80 per cent of oil reserves and 50 per cent of gas reserves, leaving the rest to the biggest traditional international companies such as ExxonMobil, Shell and BP.
30. A new constitution was adopted in 1993 that reinforces the presidential powers.
31. Federal Law on 'State Secrets', 11 November 2003.
32. The disaster, which involved two passing passenger trains, happened near the town of Asha in the Urals on 4 June 1989.
33. Gratsianov and Pimakhin (2005).
34. A group of Russian economists directed by Leonid Abalkin estimated that between $56 and $70 billion flew from the country for the sole years of 1992–3. *Guardian Weekly*, 23 May 1999.
35. In 2005, the US House of Representatives blocked the attempt of the China National Offshore Oil Corporation (CNOOC) to acquire Unocal, the US oil company, in an $18 billion bid.
36. IEA (2006: 56).
37. *Vesti*, 12 December 2007.
38. Since 2000, 1.5 per cent average annual growth in primary energy demand.
39. In 2006, gas exports outside the CIS totalled 160 bcm and 40.7 bcm to the CIS. See Campaner (2007a).

40. Ibid.
41. *Rossiyskaya Gazeta,* 10 February 2002.

References

Campaner, N. (2007a) 'The EU–Russia Energy Trade: Lessons from History', *Russia in Global Affairs*, 8 (in Russian).

Campaner, N. (2007b) 'The Eastern Vector of Russian Oil and Gas Exports', in M. Korinman and J. Laughland (eds), *Russia: a New Cold War?* Portland: Vallentine Mitchell & Co. Ltd., pp. 256–76.

Dienes, L. (2004) 'Russian Oil Prospect', *Johnson's Russia List*, 2 June.

Glaziev, S., Kara-Murza, S. and Batchikov, S. (2003) *The White Book: Economic Reforms in Russia 1991–2001*. Moscow: Istoria Rossii.

Gratsianov, L. A. and Pimakhin, A. N. (2005) 'Avarii na neftepromyslakh I magistralnykh gazoprovodakh'.

International Energy Agency (2006) *Optimising Russian Gas*. Paris: IEA.

Khlebnikov, P. (2001) *The Godfather of the Kremlin: Boris Berezovsky and the Looting of Russia*. Orlando, FL: Harcourt.

Russian Government (2003) *Energy Strategy of Russia for the Period to 2020*, 28 August.

Stern, J. (2005) *The Future of Russian Gas and Gazprom*. Oxford: Oxford University Press and OIES.

Yergin, D. (1991) *The Prize: the Epic Quest for Oil, Money, and Power*. New York: Simon & Schuster.

Zaleski, C. P. (2006) 'The Future of Nuclear Power in France, the EU and the World for the Next Quarter-Century', *Perspectives on Energy*.

4
Energy Poverty and Economic Development

Jean-Marie Chevalier and Nadia S. Ouédraogo

Energy poverty is always associated with economic poverty. It concerns people that have low income, low energy consumption and no access, or limited access, to modern energy fuel (petroleum products and electricity). Approximately 1.6 billion people do not have access to modern energy fuels. Moreover, a great number of them have no access to clean water. This means that they do not have access to economic development and they spend a good deal of their time collecting water and local energy resources such as wood and dung, leading to health problems and accelerating deforestation. Access to energy and water is an important component of the Millennium Goals.

Energy inequalities have already been mentioned. Per capita energy consumption is 0.5 ton of oil equivalent (toe) in sub-Saharan Africa, 1 toe in China, 4 in Europe and 8 in the United States. Per capita consumption of commercial energy in the United States is 80 times higher than in Africa, 40 times higher than in South Asia, 15 times higher than in East Asia, and 8 times higher than in Latin America. Parts of world are 'over-energised'; others are 'under-energised'.

Energy poverty can be defined as 'the absence of sufficient choice that allows access to adequate energy services, affordable, reliable, effective and sustainable in environmental terms to support the economic and human development' (Reddy 2000: 44). According to this definition, energy poverty is an obstacle to economic development but energy poverty is basically explained by a low income situation. The purpose of this chapter is to understand more clearly the link between energy poverty, economic poverty and environmental fragility, and then to see how access to modern energy may trigger economic development.

1 Economic poverty and energy poverty

When estimating poverty worldwide, the World Bank uses reference lines expressed in a common monetary unit for all countries. Since 2005 a new standard for measuring purchasing power parity (PPP) has been established (World Bank 2008a). The world is divided into four categories:

- High income countries: gross national income (PPP) above $11,116 per year per capita; it covers 60 countries, 1 billion people.
- Upper-middle income countries: between $3,596 and $11,115; 42 countries, 811 million people.
- Lower-middle income: between $906 and $3,595; 55 countries, 2.3 billion people.
- Low income countries: below $905; 53 countries, 2.4 billion people.

Monetary income is not by itself a perfect indicator for measuring wealth and poverty. Other factors have to be taken into account. For measuring the level of development, UNDP combines indicators of life expectancy, educational attainment and income into a composite Human Development Index (HDI). To measure poverty UNDP uses the Human Poverty Index (HPI) which uses indicators of the most basic dimensions of deprivation: a short life, lack of basic education and lack of access to public and private resources.

In this chapter we look at countries from the two lowest income categories. These categories are very heterogeneous in terms of energy endowment and energy consumption. Several groups may be identified:

- The poorest countries of the world, with a per capita income of less than two dollars per day, no oil, no gas, and a limited amount of domestic energy resources. These countries are directly concerned by the Millennium Development Goals which aim to halve the proportion of people whose income is less than $1 a day. Access to modern energy fuels and economic development are the key to reaching the targets. Among the 52 countries that are in the low income category, 37 are located in sub-Saharan Africa and South Asia.
- India and China are included in these income categories. A specific analysis has been given in Chapter 2.
- The two income categories also include some energy-rich countries. Six among the eleven OPEC countries are there: Algeria, Angola, Ecuador, Iran, Iraq and Nigeria. Other energy-rich countries, outside

OPEC, are Bolivia, Chad, Congo (Brazzaville) and Indonesia. They are poor because all of them suffer from the oil curse, a question which is analysed in Chapter 5.

This rapid description shows that, for some countries, poor access to energy is an obstacle to economic development while, for other countries, the abundance of domestic resources of hydrocarbons does not lead to economic development and may aggravate poverty and income inequalities through the oil curse mechanism.

1.1 Demography of the poor, urbanisation and structure of energy consumption

According to the United Nations' demographic forecast, the world population could reach 9 billion inhabitants by about 2050, four times what it was in 1950. Is it sustainable in terms of human needs, food and energy resources? The largest part of this increase will be in the less developed countries where the average demographic growth rate is high. This evolution shows that the world demographic, political and economic balance is shifting rapidly.

One important component of demographic evolution is urbanisation. The size of the urban population in the global population, which was 3.5 billion people in 2006, could rise to 6 billion in 2050. Each year, more than 60 million people are added to the global urban population. The extent of poverty is generally higher in rural areas than in urban areas but this may change. In Latin America, where urbanisation has been very fast, a majority of the poor now live in urban areas. Urbanisation shifts upward the demand for energy because of private and public transportation, waste management, the use of charcoal instead of wood fuel and a greater need for petroleum products (kerosene, butane, gasoline) and electricity.

1.2 Energy consumption and income

In poor countries energy consumption per capita is low, less than 1 toe in low income countries. This low consumption is further aggravated by highly inefficient patterns of consumption and production. The main characteristics of most energy systems are:

- Heavy reliance on traditional biomass (wood, agricultural residues and dung) that has negative effects on health and the environment.

- A low rate of electrification. More than 80 per cent of the people who have no access to electricity are located in South Asia and sub-Saharan Africa.
- Oil products are used for lighting (kerosene), cooking (butane), transport (gas oil and gasoline) and power generation (fuel oil, gas oil).

1.2.1 Traditional biomass

In developing countries, more than 2 billion people rely on traditional biomass for cooking and heating. Over half of the people relying on biomass live in India and China but the proportion of the population depending on biomass is much higher in sub-Saharan Africa. In East Asia, the highest proportion of people relying on biomass occurs in the Philippines, Thailand, Myanmar, Laos, Cambodia and Vietnam. In Latin America, this is the case in many Central American countries (Guatemala, Honduras, Nicaragua).

The extensive use of biomass has many negative effects on economic development, human health and the environment:

- *Waste of time*: women and children spend several hours a day collecting wood fuel and other forms of biomass. This is often to the detriment of productive activities and education. In rural sub-Saharan Africa, many women carry 20 kilograms of wood fuel an average of five kilometres *every day* (IEA 2002).
- *Human health*: the use of biomass in traditional stoves or open fireplaces is not only inefficient but also damaging for health. Combustion is incomplete and results in substantial emissions that, due to poor ventilation, produce a high indoor pollution. Exposure to this polluted air leads, among others things, to respiratory illness, cancer, tuberculosis, low birth weight and eye disease (Bruce et al. 2000). Women and children are the first to be affected.
- *Environment*: collecting wood for fuel is frequently damaging to the environment through deforestation, erosion, soil degradation and desertification. In addition, the more biomass is used for households (dung), the less is available for fertiliser.

1.2.2 Lack of access to electricity

In low income countries, the rate of electrification is low: less than 25 per cent in sub-Saharan Africa, 50 per cent in South Asia and 80 per cent in Latin America. In countries such as Burkina Faso, the Democratic Republic of Congo and Mozambique the rate is below 10 per cent. More than 80 per cent of the people who do not have access to electricity are located in South Asia and sub-Saharan Africa. Lack of electricity is correlated to income level but also to the geography and density of specific countries. In many poor countries, with high population growth, the electrification rate, which is low, is actually declining because of lack of investment and poor management and maintenance of the existing plants. Blackouts and outages are very frequent. The situation is aggravated by the rise in oil prices because many power plants are fuel-oil fired.

1.3 The financial burden of the oil import bill

All countries need oil products and, since 2004, countries have been facing an unprecedented surge in the price of crude oil and petroleum products (Bacon and Kojima 2006). Each category of petroleum products has its own specificity and demand characteristics:

- *Gasoline* is used for individual cars and small public transport vehicles (taxis, rickshaws). Gasoline consumption concerns mostly the high income fraction of the population and the administration.
- *Diesel (gas oil)* is used, in certain cases for cars, but mostly for transport (small and full-size buses, all trucks) and for agriculture (tractors and irrigation). In many countries where power outages are frequent, diesel is used as a backup fuel in small-size power generating units. The sharing between gasoline and diesel consumption depends on the technical structure of the automotive fleet and on the fiscal policy for both fuels.
- *Kerosene for aviation (jet fuel)* is also the strategic fuel for armies, not only for planes but quite often for tanks and trucks.
- *Kerosene* (outside aviation) is the fuel of choice for lighting in households that are not connected with electricity. It is much more efficient than wood fuel for cooking and heating but generally much more expensive if there are no subsidies. Kerosene, which can be sold in small individual quantities, is *par excellence* a 'social fuel'. It has been historically subsidised in many developing countries. However, since kerosene can be used as a substitute for diesel, a large price differential between the two fuels is leading to substitution and market distortions.
- *Liquefied petroleum gas (LPG)*, mainly butane and propane, are used, like kerosene, for lighting, heating and cooking. Small bottles of butane are also frequently subsidised.

- *Heavy fuel oil,* which is always the cheapest, most polluting petroleum product, is used for industry and power generation. It is an important strategic fuel, in competition with local or imported coal for power generation.

Some of these petroleum products might be replaced by substitutes, mainly biofuels and natural gas, but energy systems are very rigid and inter-fuel substitution is a long process.

A country's dependence on oil can be measured by the net oil import bill relative to GDP and its evolution over time. For many poor countries that are totally dependent on oil imports, the oil bill is an important component of the balance of payments equilibrium. These countries are vulnerable to oil price shocks.

More precisely, oil dependence and the vulnerability of countries to oil disruption or oil price shocks may be measured by a ratio which was proposed by the World Bank (UNDP/ESMAP 2005). This ratio can be understood as the product of three terms, each of which has an important signification:

$$\text{Oil imports/GDP} = (\text{oil imports/total oil use}) \times (\text{total oil use/total energy use}) \times (\text{total energy use/GDP}).$$

The ratio (oil imports/total oil use) measures the external dependence for oil consumption. This ratio can be improved by enhancing domestic exploration aiming at discoveries and production. The second ratio (total oil use/total energy use) measures the economy's dependence on oil. It can be affected by policies to encourage inter-fuel substitution or better diversification of the energy balance. The third ratio (total energy use/GDP) measures energy intensity. It can be improved by increasing energy efficiency and also by a shift in the structure of production from energy-intensive activities to less energy-intensive sectors. Such changes imply structural measures that reflect a medium- or long-term vision of the energy future for a given country.

On a shorter-term perspective, the rise in prices of oil and petroleum products that has occurred since the beginning of 2004 is having devastating effects on certain countries and certain customers. The oil bill has reached a level of 15–20 per cent of GDP in certain poor nations. Governments are under pressure to take measures to alleviate the burden. ESMAP (the World Bank Group) has conducted a survey in 38 developing countries to evaluate the solutions that have been found by governments for coping with higher oil prices (Bacon and Kojima 2006).

Table 4.1 Retail gasoline and diesel prices per litre in US$ cents (November 2006)

	Gasoline	*Diesel*
United Kingdom	163	173
France	148	133
United States (Benchmark)[1]	63	69
Indonesia	57	44
Malaysia	53	40
Nigeria	51	66
Angola	50	36
Algeria	32	19
Saudi Arabia	16	7
Libya	13	13
Iran	9	3
Venezuela	3	2

[1]Green Benchmark reflects a market price taking into account infrastructure expenditures.
Source: GTZ, International Fuel Prices (2007).

Political solutions from governments include price-based policies (full or partial pass-through of higher prices, subsidies, tax adjustments, compensation funds), quantity-based policies (rationing, mandatory conservation) and structural policies (fuel switching, energy efficiency).

Among the political solutions, the question of subsidies is one of the most important. A number of countries have been convinced in the past to establish subsidies for certain energy products. This is particularly the case of oil exporting countries which use subsidies to transfer part of the oil rent to the population (Table 4.1). Venezuela is an extreme example. Iran, which has some of the lowest prices, is one of the most energy-intensive economies in the world. But oil importing countries are also establishing subsidies that pose some problems when prices rise because it increases the gap between domestic capped prices and international market prices. This is the case in China, India, Morocco and many other countries. When people are accustomed to low subsidised prices, they are likely to resist more strongly any increase to the international market level.

Indonesia provides an interesting and painful example. Indonesia is a major oil producer and was long a major oil exporter. This situation led governments to set up domestic prices for petroleum products well below international prices. This induced widespread smuggling of subsidised products out of the country. Then, in 2001, oil production began to decline, partly due to the lack of investment. At the same time domestic demand increased drastically, encouraged by low prices and a strong demographic growth. In 2004, the country became a net oil importer. The year 2004 corresponds to the beginning of the upsurge in crude oil prices in international markets. The government had no choice but to try to progressively remove subsidies. Two substantial price increases were made in 2005. These two increases still left gasoline and diesel prices at 80 per cent of international market prices and kerosene at 40 per cent. However, in contrast with the other adjustments made in the past which had led to violent opposition, the price increases of 2005 were accepted, thanks to the credibility of the newly elected government. At the same time, the parliament established a cap on the total amount of subsidy included in the budget. However, the cap rapidly became politically unsustainable because of the continuing rise in oil prices and the continuing decline in domestic oil production. In 2008, petroleum products subsidies represented 22 per cent of state expenditures, more than the cumulated budgets for education, health and infrastructures (*Le Monde*, 30 May 2008). New price increases were decided in May 2008 (about 25 per cent) leading to violent popular opposition. At the same time Indonesia decided to quit OPEC.

The question of subsidies (for oil products, electricity and natural gas) illustrates the fact that energy prices are politically sensitive matters. When international prices go up, it puts political pressure on governments and the cost of alleviation is very high. In addition, price subsidies tend to benefit mostly the rich since their level of consumption is higher. In Morocco, it has been estimated that 75 per cent of subsidies for petroleum products go to 20 per cent of the richest people. Cross subsidies between products are possible (kerosene is used more by the poor) but their management induces high transaction costs, especially because of smuggling (Aoun 2008).

1.4 Climate change vulnerability

Most of the studies concerning climate change and its consequences show that developing countries may be the first victims of climate change through a number of phenomena such as soil erosion, floods, droughts, wind storms, extreme climate events and pollution. Furthermore, their reliance on agriculture and their low income situation make it more difficult for them to adapt. In sub-Saharan Africa, the number and impact of natural disasters are both increasing, affecting millions of people every year.

Table 4.2 Countries most at risk from climate-related threats

Drought	*Flood*	*Storm*	*Coastal 1m*[a] *All low-lying*	*Coastal 5m*[a] *All low-lying*	*Agriculture*
Malawi	Bangladesh	Philippines	Island States	Island States	Sudan
Ethiopia	China	Bangladesh	Viet Nam	Netherlands	Senegal
Zimbabwe	India	Madagascar	Egypt	Japan	Zimbabwe
India	Cambodia	Viet Nam	Tunisia	Bangladesh	Mali
Mozambique	Mozambique	Moldova[b]	Indonesia	Philippines	Zambia
Niger	Laos	Mongolia[b]	Mauritania	Egypt	Morocco
Mauritania	Pakistan	Haiti	China	Brazil	Niger
Eritrea	Sri Lanka	Samoa	Mexico	Venezuela	India
Sudan	Thailand	Tonga	Myanmar	Senegal	Malawi
Chad	Viet Nam	China	Bangladesh	Fiji	Algeria
Kenya	Benin	Honduras	Senegal	Viet Nam	Ethiopia
Iran	Rwanda	Fiji	Libya	Denmark	Pakistan

Note: The typology is based on both absolute effects (that is, total number of people affected) and relative effects (that is, the number affected as a share of GDP). [a]Metres above the sea level. [b]Winter storms.
Source: World Bank (2008d).

Various organisations predict some measure of climate change vulnerability. Table 4.2 gives a list of countries that are most at risk from climate change threats. Among the listed countries, 25 belong to the low income category.

The University of Oxford has built a *Climate Vulnerability Index* which, among other measures of vulnerability, expresses the vulnerability of human communities in relation to water resources (other indexes are available; see Vincent 2004). At present a great number of climate change consequences have to do with water: lack of water or an excess of water. The scores of the index range on a scale of 0 to 100, with the total being generated as a weighted average of six major components:

- The physical availability of water resources.
- The quality of access to resources.
- The effectiveness of people's ability to manage water.
- The ways in which water is used.
- Ecological integrity related to water.
- Geographical specificities of the area.

All these components are interrelated. They relate to the complexity of energy issues by taking into account climate change, the global crisis in food, water and sanitation and the governance of the country. UNDP mentions that each year, sub-Saharan Africa loses more in productivity through poor water management (one of the components) than it gains through development aid and debt relief: a staggering $30 billion.

The results of the survey for the *Climate Vulnerability Index* show that the most vulnerable countries are precisely in the low income and the lower-middle income categories concerned in this chapter. Vulnerability of the poor countries means that it will be more difficult and more costly for them to mitigate and to adapt to the threats of climate change. It could aggravate tensions, migrations and wars. Climate change is bringing a new dimension to the geopolitics of the planet. The war in Darfur is a complex political story but it has a climate change component related to drought. Elsewhere in the world, the first climate change refugees appeared in the Pacific islands in 2008.

1.5 Deforestation: one of the main causes of climate change

Deforestation is responsible for more than 18 per cent of global greenhouse gas emissions, more than that emitted by the global transport sector. Forest assets are not distributed evenly across the world. Low and middle income countries account for almost 80 per cent of the world's forest area. Eight out of the top ten countries in terms of annual deforestation (Table 4.3) belong to the low and lower-middle income countries. Most of this deforestation concerns rainforests.

Deforestation is the result of conflicts of interest between short-term private interests, government activities that favour or limit deforestation,

Table 4.3 Annual deforestation for the top ten countries

Countries	*Thousands of square km/year 2000–5*
Brazil	31
Indonesia	18.7
Sudan	5.9
Myanmar	4.7
Zambia	4.4
Tanzania	4.1
Nigeria	4.1
Congo, Dem. Rep. of	3.2
Zimbabwe	3.1
Venezuela, R.B. de	2.9

Source: CGEMP based on World Bank (2008c).

pure national interest and global welfare aiming to stabilise world emissions and preserve biodiversity. Deforestation has a number of causes. In addition to the traditional collection of wood for cooking, deforestation is mostly due to short-term money seeking conduct.

- *The use of wood fuel* may accelerate deforestation in certain areas. In addition, the process is amplified when accessible forests are close to cities. Then the wood is transformed into charcoal to be sold in the urban markets where there is a high demand, especially when the price of oil products (kerosene and butane) is high. Illegal charcoal trade has devastating effects in the Congo basin rainforest.
- *The market for cheap pulpwood* is a very active market for various industries: construction, furniture, pulp and paper manufacturing.
- *The extension of traditional agricultural production.* Slash-and-burn techniques are used to clear forests and clear open space for planting large single crops (bananas, palms, manioc, maize, soybeans, cacao). The productivity of the soil declines after a year or two and farmers move to clear other areas.
- *The extension of modern large-scale production.* Forests are also opened up for modern large-scale agriculture or cattle-raising aimed at export markets. The production of low quality cheap meat is an attractive activity driven by the global fast food industry. In Brazil, the development of sugar cane production for ethanol accelerates the opening of new territories in forest zones for cattle breeding and soybeans.
- *The extension of infrastructures.* Urbanisation, the development of mining activities and large industrial projects, including hydroelectric projects, call for large extensions of infrastructures that are another cause of deforestation.

Deforestation poses many problems. At the local level, continuing deforestation may increase a country's vulnerability to climate change. At a global level, continuing deforestation weakens our possibility of resistance. This was acknowledged at the Bali conference (December 2007). Research carried out for the Stern Review indicates that the opportunity cost of forest protection in eight countries responsible for 70 per cent of emissions from land use could be about $5 billion per annum initially, although, over time, the marginal cost would rise (Stern Report 2006). At the Bali conference, propositions were made to put in place a fund to compensate for the Reduction of Emissions from Deforestation and Degradation (REDD). The problem is now on the international agenda.

2 Energy for economic development

Economic poverty is linked with energy poverty and, at the same time, energy is an important vector for triggering economic development and for reaching the objectives of the Millennium Goals. The quantitative relationship between energy consumption and economic growth is a matter of debate (Keppler 2007), but qualitatively, the role of energy in economic development is clearly indicated in Table 4.4. In brief, access to modern energy sources is a condition for:

- setting up new productive activities, generating jobs, saving time for work and education;
- increasing business opportunities through better and faster communication systems;
- improving access to clean water;
- improving health, education and nutrition;
- protecting the environment;
- promoting gender equality;
- enlarging opportunities for international partnerships.

The process of 'energy transition' shows that within the dynamic of economic development, poor families in developing countries gradually increase their revenues and get access to modern energy fuels, basically electricity and petroleum products. Access to modern energy fuels is a prerequisite for economic development but other conditions are also required: access to clean water, education, health services, and infrastructure for transport and telecommunications. To develop access to energy in a given country, the first step is to evaluate local energy resources and their potential. Another priority is 'electricity for all' which enables access to mobile phones, TV and the whole communication network including the internet.

2.1 Energy resources

The paradox of countries where people have limited access to modern fuels is that many of them have a huge potential of untapped renewable resources such as biomass, sun, wind and rivers. In addition, there is an important domestic resource with a huge potential: the improvement of energy efficiency. In some countries, this potential is progressively being exploited but there are a number of obstacles to overcome: financing the investment (public vs. private), mitigating the risks, managing the projects, ensuring maintenance, and mobilising the appropriate human

Table 4.4 Energy and Millennium Development Goals linkages

MDG	Energy Linkages
1 Eradicate extreme poverty and hunger	Energy inputs such as electricity and fuels are essential to generate jobs, industrial activities, transportation, commerce, micro-enterprises, and agriculture outputs. Most staple foods must be processed, conserved and cooked, requiring energy from various fuels.
2 Achieve universal primary education	To attract teachers to rural areas electricity is needed for homes and schools. After dusk study requires illumination. Many children, especially girls, do not attend primary schools in order to carry wood and water to meet family subsistence needs.
3 Promote gender equality and empower women	Lack of access to modern fuels and electricity contributes to gender inequality. Women are responsible for most household cooking and water-boiling activities. This takes time away from other productive activities as well as from educational and social participation. Access to modern fuels eases women's domestic burden and allows them to pursue educational, economic and other opportunities.
4 Reduce child mortality	Diseases caused by unboiled water, and respiratory illness caused by the effects of indoor air pollution from traditional fuels and stoves, directly contribute to infant and child disease and mortality.
5 Improve maternal health	Women are disproportionately affected by indoor air pollution and water- and food-borne illnesses. Lack of electricity in health clinics, lack of illumination for nighttime deliveries, and the daily drudgery and physical burden of fuel collection and transport all contribute to poor maternal health conditions, especially in rural areas.
6 Combat HIV/AIDs, malaria, and other diseases	Electricity for communication such as radio and television can spread important public health information to combat deadly diseases. Health care facilities, doctors and nurses all require electricity and the services that it provides (illumination, refrigeration, sterilisation, etc.) to deliver effective health services.
7 Ensure environmental sustainability	Energy production, distribution and consumption have many adverse effects on the local, regional and global environment; these effects include indoor, local and regional air pollution; local particulates; land degradation; acidification of land and water; and climate change. Cleaner energy systems are needed to address all of these effects and to contribute to environmental sustainability.
8 Develop a global partnership for development	The World Summit for Sustainable Development (WSSD) called for partnerships between public entities, development agencies, civil society and the private sector to support sustainable development, including the delivery of affordable, reliable and environmentally sustainable energy services.

Source: United Nations.

resources. High oil prices are increasing the economic attractiveness of these developments but the general institutional and educational environment has to be improved substantially. However, we will see later on that decentralised private initiatives could be undertaken in these areas. Let us focus on biomass and energy efficiency.

2.1.1 Biomass

Biomass resources fall into two different categories: (i) traditional biomass: the fuel of the poor with its associated negative effects that have been described above and (ii) biomass to be transformed into biofuels for complementing or replacing petroleum products. First-generation biofuels include bioethanol, which is derived from sugarcane, corn or sugar beets, and biodiesel, which is produced from oil crops such as rapeseed, soybeans or palm (see Box 9.1 in Chapter 9). Brazil has been a pioneer in developing ethanol from sugarcane as a substitute for gasoline. Thailand has followed the same path. Indonesia and Malaysia have chosen to develop palm oil on a large scale, a choice which clearly contributes heavily to the destruction of rich bio-diverse rainforests.

Since the surge in oil prices, a number of developed and developing countries are considering the possible development of biofuels. Brazil is proposing its technologies to other developing countries in Central America and Africa. The development of biofuels requires very careful analysis on a local basis also because a national decision may have an international impact on the energy/environment issue. Some of the problems that have to be examined carefully are:

- On a local basis, the economic cost of producing biofuels and its expected evolution have to be precisely evaluated. The tax impact has to be calculated by comparing biofuel taxation (or subsidies) with the tax take on petroleum products.
- On the same local basis, the environmental impact must be evaluated: GHG emissions, taking into account the consumption of fertiliser, and the impact on deforestation and land use. Certain African countries (Ghana, Tanzania) might have an important potential for the production of biofuels on land that cannot be used to produce food.
- On a global basis, the massive development of biofuels could have a negative impact on GHG emissions if fertilisers, deforestation, change in land use and other components are taken into account. It could also impact the price of sugar and other agricultural products. In 2008, the World Bank reported that the development of biofuels in the United States and Europe bears some responsibility for the increase of prices of food products between 2002 and 2008.

Box 4.1 Brazil: an ethanol champion

Due to its rich endowment in natural resources, its geographical position and its policy of energy independence, Brazil holds a very exceptional position since the contribution of renewable energy to primary energy consumption reaches 46 per cent. Part comes from hydropower (85 per cent of electricity generation comes from hydro facilities) and the other part comes from ethanol used as a complement or a substitute for motor gasoline. The decision to develop ethanol production from sugarcane was taken in the 1970s. With oil imports standing at 80 per cent at that time, the country was deeply affected by the oil crisis. Today, most cars now sold in Brazil are able to switch between 100 per cent ethanol and a gasoline blend. Half of the transport in Brazil is ethanol-based. However, there is a debate about the relationship which may exist between the increase of ethanol production and its impact on deforestation.

The case of biofuels confirms the growing interdependence between energy policy, climate change issues and food production and also between national political decisions and the global management of climate change. Once more we see conflicting interests between nations, between private goods and one single public good which is the climate.

2.1.2 Other energy resources

Setting aside domestic resources of oil and gas that are generally associated with the problem of the 'resource curse', developing countries have a great potential for the development of renewable energies: wind, solar and hydropower. While Europe and North America have already exploited about 60 per cent of their hydro potential, it is estimated that the rate is about 20 per cent for Asia and South America and only 7 per cent for Africa. The geographical location of these potential resources remains an obstacle for project development.

2.1.3 Energy efficiency

Another very important domestic resource to be considered is energy efficiency. In all countries, there are huge potentials for improving energy efficiency. This means that a given energy service can be met with lower energy consumption. Improving energy efficiency is a win-win strategy: it reduces the energy bill, it reduces emissions and it is 25–50 per cent cheaper than producing the amount of energy which is saved.

Vast potentials for energy saving opportunities remain untapped even though current financial returns are strong. The World Bank has been very active in the field of proposing and financing energy efficiency programmes in developing countries. The Bank has carefully reviewed the means by which energy efficiency might be improved as well as the existing obstacles. The organisation proposes 'a general model for successful delivery programs for energy efficiency investments' (World Bank 2008d). The Bank's study shows that investments in energy efficiency may be promoted through three channels:

- Loan financing schemes proposed by commercial banks or special agencies.
- Energy service companies (ESCO) that propose energy performance contracts by which the initial investment is repaid by energy savings.
- Demand side management (DSM) strategies that are set up by local utilities to encourage energy savings.

In developing countries some successful programmes have been put in place but, very often, the institutional and political environment is seen as an obstacle to a real policy of energy efficiency. The political willingness, which is crucial in these matters, is frequently lacking. The banking community is not familiar with energy issues and this sort of investment is not a priority. The legal framework is often insufficient to secure some unusual contractual arrangements. However, high energy prices could trigger the financial interest of governments for such programmes. In addition, the investments related to the Clean Development Mechanism (CDM; see Box 9.5 in Chapter 9) could contribute to sustainable economic development.

CDM projects in developing countries are mostly concentrated in emerging economies such as China and India. They both represent more than half (53 per cent) of the registered CDM activities in the world. An additional 33.3 per cent of the total CDM projects are located in upper-middle economies such as Brazil, Mexico, Malaysia and Chile or in high-income economies. Only 13.7 per cent of the CDM projects (in 2008) go to the low and low-middle income countries. These figures may evolve in coming years, with the improvement of private–public partnerships, a greater involvement of international organisations and the Reduction of Emissions from Deforestation and Degradation (REDD) initiative.

Box 4.2 Energy efficiency: the case of Tunisia

Tunisia was a net exporter of crude oil in the 1980s, but production has been declining and the country became a net importer in 2000. To manage this difficult but expected change, the government understood rapidly that the first weapon was the improvement of energy efficiency. A state agency for the management of energy demand, one of the first in developing countries, was created in 1985 (the Agence de Maîtrise de l'Energie which became the Agence Nationale de l'Energie (ANE)) which set up a series of programmes for enhancing energy efficiency and increasing the contribution of renewable energy (mainly solar). A system of subsidies was put in place for energy audits and priority investments. Between 1985 and 1992 Tunisia managed to improve its energy efficiency by 20 per cent. Since then, pressure for energy efficiency and energy savings for buildings, equipment, transport and appliances has been constantly maintained in order to improve performance. Since 2002, ANE has launched important programmes for replacing current conventional bulbs by high efficiency bulbs for public lighting in the streets and for households. In a context of high energy prices these actions are highly beneficial.

2.2 Access to electricity

Access to electricity is a prerequisite for economic and social development. It has been noted already that 1.6 billion people are still not connected to electricity. New connections are made by grid extension (for urban, peri-urban and rural households) or through the implementation of decentralised power systems using local resources (hydro, solar, biomass, wind) or diesel generators. In the IEA's reference scenario, if no new policies are put in place, there will still be 1.4 billion people not connected in 2030. The rate of electrification is still very low in poor countries (see above) and progress is slow. In a general review of rural electrification programmes, the World Bank (2008c) considers that, generally speaking, off-grid systems are more expensive and bring fewer benefits than grid connections. Photovoltaic cells, for example, are often limited in capacity and offer fewer hours of lighting. In addition, decentralised power systems often face operational and maintenance problems due to a lack of skilled human resources. However, the combination of high oil prices with progress in cost and efficiency of renewable energy systems (batteries) could enhance the attractiveness of decentralised systems in areas difficult to reach with the grid. Large programmes of decentralised electrification, through solar systems and bio-digesters, have been developed successfully in Bangladesh (Yunus 2007).

The 'willingness to pay' for consuming electricity is generally higher than the long-term marginal cost of supply but there is the problem of payments of connecting charges that have sometimes to be extended over a long period of time. This means that, in many rural electrification programmes, access to electricity is not brought to the very poor.

Access to electricity has major impacts on daily life. Benefits concern comfort, health, education and local production.

Comfort: In poor rural areas, when a village is connected, the major primary uses of electricity are lighting, television and mobile phones. Electric lighting increases the time available for education or productive activities. More light means that people may work longer, generating higher income. TV brings information about the global village. After lighting, TV, radio and mobile phones, there is a progressive transition towards higher standards of living if there is economic growth: the acquisition of refrigerators (which provide better food conservation), fans, irons, small appliances and air conditioning.

Health: Kerosene is the main fuel for lighting when there is no electricity. Kerosene is expensive (even when subsidised), polluting and inefficient. It is, with wood fuel, the main element responsible for indoor pollution. Moving from kerosene to electricity significantly cuts the cost of lighting, with much greater luminosity and much less pollution. Rural electrification also transforms the working conditions of local clinics and health centres which can run for longer periods and use more equipment. However, the cold chain for vaccines and medicines may remain fragile because of the risk of power supply disruptions that are very frequent in developing countries. Electricity may also facilitate access to clean water (pumping).

Education: Various surveys show that children in electrified households have higher education levels than those without electricity. They are able to spend more time studying (at home and at school), they have access to information (especially through the internet) and their village becomes more attractive for teachers (and doctors).

Production: Electricity enlarges business opportunities and extends working hours. Electricity may be used for increasing the productivity of agriculture (pumping and irrigation), and of industry and services. The most important potential transformation is probably for small businesses, including home enterprises. According to the World Bank,

Box 4.3 The case of rural electrification in Morocco

In 1995, the government decided to launch a programme of rural electrification aiming to raise the rate of rural electrification from 26 per cent to more than 95 per cent. The state-owned power company, ONE, was responsible for the programme. Two types of electrification were undertaken: grid extension and connection for 16,500 villages, and photovoltaic decentralised systems for 3,500 villages. For decentralised systems ONE set up a system of public–private partnerships where ONE granted ten-year concessions to private companies for the installation and management of those systems. The financing of the programme (about €2 billion) was set up as a combination of contributions from ONE, local communities, new customers benefiting from the programme and a tax on all other consumers. Some loans came from foreign public agencies such as the Agence Française de Développement. The organisation of the project, with a strong governmental commitment, has been very successful. A rate of 97 per cent for rural electrification was achieved in 2007. Two million households have benefited from the programme with clear effects on poverty reduction and the creation of new business opportunities.

Source: Butin and de Gromard (2006).

for example, women in Ghana prepare snacks to be sold to the people who come to their houses to watch TV in the evenings. In South Africa, households sell cold drinks and rent out refrigerator space. We will see below what could be the impact of combining electricity with the new technologies of information and communication.

2.3 Energy and economic development through ICT

Connection to electricity opens the door for the use of information and communication technologies (ICTs). A number of research studies have been undertaken to measure the impact of ICTs on economic development. Quantification is difficult because case studies are scattered but there is general agreement that ICTs play, and will increasingly play, an important role in economic and social development and poverty reduction. When electricity is available, the two major applications of ICTs are mobile phones and the internet. However, access to the internet depends on the level of education. The main benefits of ICTs in developing countries are access to information, access to markets and access to credit.

Box 4.4 Example of telemedicine in Gambia

In Gambia, nurses working in remote rural hospitals are using digital cameras and laptops to send photographs and describe the symptoms they do not recognise and/or cannot handle. The images are transferred to a doctor in Banjul who makes the diagnosis. The doctor himself may ask for help from a British health agency.

Source: International Telecommunication Union (1999).

Box 4.5 The Grameen Phone

The Grameen Phone, in collaboration with the Grameen Bank (founded by Mohammad Yunus, Nobel Peace Prize winner) lends money to villagers in Bangladesh (often women) who buy a mobile phone. In exchange for payment, they then grant telephone access to their entourage. A study by the Canadian Agency for International Development reported that this practice is a success, each phone generating an average income of over US$100. In addition, the system improves the living standards of villagers.

Source: Hammond (2001).

These factors may impact local business and the range of opportunities for creating new businesses.

Access to information: Mobile phones and the internet considerably improve access to information. It may concern agricultural and irrigation techniques, weather forecasting or the appropriate use of fertiliser. ICTs provide great potential for education, learning and training. New technologies may cheaply enhance the educational system when appropriate human resources are available. In sub-Saharan Africa, twenty-two countries are currently linked to the 'Virtual African University'. Students interact directly with faculty members and have access to an online library which also provides courses in computer science, economics and languages. For health, access to information is vital for rural hospitals and health centres. 'Telemedicine' is developing by the intensive use of ICTs.

Access to markets: Mobile phones and the internet facilitate contact with markets. In Mauritania, fishermen follow the prices on the fish market and know when they have to stop fishing. The internet also offers the possibility of expanding markets, broadening the range of customers

Box 4.6 Telecentres in Senegal

Telecentres were developed in an independent way through contracts with the main telecom operator, Sonatel. There are about 10,000 privately owned telecentres (including 4,000 located in rural areas and small towns) employing about 18,000 people with a revenue of about US$800 per year for each person, which is well above the minimum salary in Senegal. The telecentres have created twice as many jobs as Sonatel alone. Furthermore, the centres have created an infrastructure offering an excellent basis for providing villages and small towns with collective access to the internet. This is a good example of how information technologies are also providing opportunities for job creation and development in poor countries. People who do not have home telephones can easily use telecentre facilities.

Source: UN Newsletter, November 2000.

and facilitating customs procedures, transport and logistics. Local companies may have access to global markets.

Access to credit: ICTs facilitate the access to microcredit and microfinance for business creation. Microcredit aims at lending to the poor, who have no access to the banking system, small amounts of money ($30–$100), with no guarantees, to help them create an economic activity. It was introduced in Bangladesh by Mohammad Yunus on the principle that 'The poor always pay back'. It is now promoted in many countries by the Grameen Trust (Yunus 2007). From the Grameen model a number of new, more efficient, business models have been launched in many poor countries (Joubert 2006a, 2006b). One may also mention PlaNet Finance created by Jacques Attali which 'aims to bring the full potential of the internet for development of microcredit' and Kiva.org, a microcredit institution set up by students from Stanford.

The International Labour Organization notes that the spread of ICTs in developing countries creates millions of jobs of many different kinds (Curtain 2001). These include:

- the people directly employed in telecentres, internet cafés and other 'shops' linked to the internet and mobile phones;
- the jobs of 'coaching' for computer training, assistance, maintenance tools, repair equipment and the sale of telephone cards;
- jobs found through the internet or businesses that have expanded the scope of their activities through ICTs.

3 Energy and economic development: a geopolitical approach

In developing countries, access to energy and to economic development relies heavily on state support and government commitment. It is government's responsibility to establish a clear institutional framework and to decide the role that is to be given to state-owned companies, private national capital and international investors.

3.1 The nature of the state

In the countries under review in this chapter, the nature of the state varies widely from one country to another. It is very difficult to establish a classification of political systems but extensive research has been conducted to measure the nature and quality of governance and its relationship with economic development. The World Bank has established an index of governance where six components of governance are taken into account (Kaufmann et al. 2007):

- *Voice and accountability* which measures citizens' participation in the choice of governments, and freedom of expression.
- *Political stability* which measures the risk of violent political destabilisation.
- *Government effectiveness* which measures the quality of public service, the absence of sensitivity to political pressure and government's credibility.
- *Regulation quality* which measures government's ability to establish a stable legal and institutional framework.
- *Rule of law* that protects human beings and contractual arrangements and guarantees legal settlements.
- *Control of corruption* which measures how corruption is evaluated by various entities doing business in the countries.

A number of other indicators have also been established. They concern democracy, transparency, corruption, country risk and business conditions. Another geopolitical approach is based upon ethnic fractionalisation. Ethnic fractionalisation can be defined as the probability that two persons randomly drawn from a given society are from the same ethnic group. Many countries of the lower income categories suffer from ethnic fractionalisation, particularly in Africa, Asia and Latin America. Sub-Saharan countries provide a large diversity of ethnic fractionalisation with very often one majority group and the rest of the population

Box 4.7 Nigeria: an illustrative case for the oil curse

Poverty in the midst of abundance is a well-known paradox characterising the Nigerian economy. Nigeria is a nation blessed with abundant oil resources. It is ranked as the sixth largest exporter of petroleum in the world with annual oil revenues of US$47 billion (2005) representing 47 per cent of its GDP.

However, despite its bountiful oil resources and cumulative revenue, Nigeria's economic performance has been startlingly poor. Between 1970 and 2000, the number of people living on less than one dollar per day rose from 36 per cent to 70 per cent. Income inequality has widened considerably. Between 1980 and 2000 per capita income decreased from US$1,021 to US$878. Indexes of governance and human development are low. Moreover, the index of ethnic fractionalisation is one of the highest. All these elements explain the persisting high political and social tensions in the producing area of the Niger delta. Many groups of various natures are fighting against each other and against the oil companies to capture part of the oil money.

Despite this situation, in 2004 Nigeria signed the Extractive Industry Transparency Initiative (see Chapter 9) in order to improve the transparency of financial flows related to the oil industry.

divided among quite small groups. The effects of ethnic fractionalisation have been extensively studied (Alesina et al. 2003). Highly fractionalised societies face a greater level of political competition between rent-seeking groups, resulting in higher transaction costs to reach an agreement on public goods like health, education and infrastructure. Economic growth and wealth from natural resources are unevenly distributed. Each ethnic group prefers targeted goods from which they get primary benefits rather than public goods whose gains are shared with other groups.

All these analyses concerning the nature of the state show that the relationship between energy resources, energy policy and economic development is heavily determined by the type of political governance that is in place. To take one extreme example, the energy poverty and the economic poverty of Nigeria are directly related to the nature of governance which prevails in the country. This is a perfect illustration of the oil curse.

In certain regions, there is also the danger of 'failed states': those unable to meet their responsibilities and giving way to warlords, traffickers and terrorists. The failure to build peaceful democracies in Afghanistan or Iraq and the failure to prevent some African nations from disintegrating are examples of these obstacles to economic development.

Beyond the nature of the state itself, international investors advocate a strong and stable state, even if it is a single-party (or a single individual, or a single family) government that remains in power for a long period of time. What is expected first from such governments is the creation of a clear and stable institutional framework.

3.2 The institutional framework

The institutional framework covers many elements in the field of energy and climate change. First, there are a number of important laws (petroleum law, electricity law, energy law, fiscal regimes and rules for protecting the environment) which provide general orientations and protections. This legal framework is expected to be stable. Any unexpected change may have disastrous effects on the energy sector. In Venezuela and Bolivia, the radical changes initiated in the early 2000s were devastating for the oil and gas industries and for their future. The legal framework is particularly important for property rights and the security of contractual arrangements which are numerous in the energy industry.

The efficiency of the administration is also a key concern for delivering multiple authorisations and permits, giving approvals and answering questions. Administrations are frequently burdened by bureaucracy, low standards of organisation, lack of motivation and corruption. This results in very high transaction costs and delays in projects. If the government is changed, certain projects may have to start all over again.

The organisation of the energy sector is an important matter. What should be conducted by state-owned companies? What should be left to the national and international private sector? In the energy sector the role of the public sector has generally been very strong. The typical situation is that of a state-owned vertically integrated monopoly for electricity generation, transmission and distribution. This model, which still exists in many countries of the South, was widely criticised in the early 1990s. The wind of liberalisation which blew from the United States and the United Kingdom suggested the dismantlement of the vertically integrated value chain in order to separate what was natural monopoly (the wires of power transmission and distribution) and competitive activities (power generation and power supply). It was suggested that these principles (see Chapter 1) should be applied in developing countries. The main arguments were the poor performances of state companies, states' inability to finance new investments and the need to attract international investors. Baumol and Lee (1991) wrote an innovative article which illustrated the theory of contestable markets. The purpose of the article was to demonstrate how the liberalisation of the power industry in Nigeria could improve the performance of the electrical system. The theoretical approach was new but its application proved to be difficult. After years of experimentation, the countries of the South are now more prudent about radical reforms. They accept, for the power sector, the importance of the participation of international investors. Independent power producers (IPPs) have been established in many countries. Their profitability is usually based upon long-term purchasing power agreements (PPAs) where the dominant power utility (most frequently state-owned) buys power at an agreed price. Governments are now open to private–public partnerships in order to broaden the range of financial resources. Many power systems are considered as 'hybrid'. They are not fully liberalised, however, and state-owned companies still have an important role.

3.3 International investors

Developing countries are attracting a huge flow of foreign direct investments. Despite country risks, investors are primarily interested in access to natural resources: oil and gas, coal, uranium, wood and mining products. They are also interested in the infrastructures that have to be built in these countries under demographic pressure. All sorts of investors are competing: multinational companies, but also equipment suppliers, banks and various types of funds (from private equity to hedge funds and sovereign funds). These investors proceed with the traditional approach: risk identification, risk analysis and risk mitigation (see Chapter 1).

Among foreign investors, one may notice the fast-growing intervention of Chinese and Indian entities. Everywhere in the world where there is oil, Chinese companies are there. Developing countries offer immense opportunities for China's external growth. Chinese companies' strategies are closely linked and supported by their government. They are able to propose 'global packages' to developing countries which may combine an oil, gas or coal investment combined with the building of roads, railways, power plants, refineries, hospitals and schools – and some luxurious private homes for the ruling class (Michel and Beuret 2008). In addition, the Chinese government may propose loans and financial facilities to local governments. The Chinese way of doing business tends to fit the governance structure of developing countries. This is particularly true in Africa. Chinese firms are integrated into global business systems that combine various integrated value chains. This favours 'network trading' (Broadman 2008) in which local capitalism is closely associated with Chinese businessmen. Chinese entities bring new forms of aggressive competition in which corruption, civil rights and protection of the environment are not fundamental issues. This strategy is paying off: relationships between China and oil-exporting African countries have been established on a broad interdependent basis. Angola, Sudan, Nigeria and the Republic of Congo have become China's major oil suppliers. In Sudan, Chinese companies have taken the place of multinational oil companies.

4 Reaching the Millennium Goals?

The central messages of the 2008 Global Monitoring Report on the Millennium Development Goals are clear: 'On current trends, the human development MDGs are unlikely to be met ... Progress toward MDGs is slowest in fragile states, even negative on some goals.' The diagnosis is still more pessimistic in the 2008 UNCTAD Report. As the IEA's chief economist points out: 'These prospects are unacceptable – morally, economically and politically. That is why decisive policy action is needed urgently to accelerate energy development in poor countries as part of the broader process of human development' (Birol 2007: 1).

The goals are unlikely to be met because there is not enough money and not enough investment and also because of bureaucracy, poor governance and lack of skilled resources. Indeed, developing countries suffer from many sorts of vulnerabilities and, in addition, they now face a great number of uncertainties concerning the effects of climate change and the evolution of prices for food and energy. All these difficulties are much greater for them than for rich countries.

4.1 Vulnerabilities and uncertainties

This chapter has shown that poor countries are more vulnerable to the effects of climate change, which could be devastating for a number of them, than are rich countries. This vulnerability is exacerbated by the current growth of population that increases the need for food, water, energy and infrastructure.

The increasing demand for food in developing countries poses the question of a 'global food crisis', a spectre that has haunted the world since 2007. High food prices have particularly hit vulnerable populations that spend a large amount of their income on food. Poor urban populations have been hit hard leading to the wave of riots, demonstrations and strikes which took place in more than forty countries in 2008. According to the UN FAO (May 2008), the food import bill of the Low Income Food Deficit Countries was expected to reach US$169 billion in 2008, 40 per cent more than in 2007. Security of food supply is at risk. At the same time, the same people have faced a rocketing escalation of energy prices.

The measures that have been taken by governments to alleviate the burden (fiscal measures, price control, subsidies and restrictions) are political short-term measures that may give temporary respite but will exacerbate long-term problems if the situation continues. Price increases hit the very poor. They also tend to aggravate income inequalities and to exacerbate social tensions.

4.2 Meeting the challenges

Despite all the vulnerabilities, uncertainties and difficulties, all possible actions and efforts should be mobilised to accelerate the economic and sustainable development of poor countries. The World Bank and other UN organisations play an important role in launching programmes and projects. They integrate environmental issues into national and local development plans. They play the role of catalyst for mobilising financial resources and mitigating project risks. In addition to the actions of large organisations, we believe that new forms of capitalism will play an important and growing role in economic development. From a global energy and environmental point of view, two issues are dominant: the access to electricity and climate change mitigation.

Electricity for all: Access to electricity symbolises the improvement of daily life through lighting, access to information, communication and new business opportunities. Access to electricity is developed through governmental programmes, with the support of international organisations, but during the past few years there has been a multiplication of private initiatives or private–public partnerships which focus on small local projects that favour decentralised economic development by combining access to electricity, clean technologies and business initiatives. Some examples are the Small Scale Sustainable Infrastructure Development Fund (S3 IDF), the African Rural Energy Enterprises scheme (AREED) and the Sustainable Energy Finance Initiative and Share the World's Resources scheme (see their web sites for more information).

Climate change mitigation: Climate change mitigation is another priority which has a number of dimensions. The development of renewable energies and the improvement of energy efficiency are two major components of a national energy policy. At a more general level, developing countries face climate change issues. Some of them will have to adapt and the cost of adaptation might be very high in terms of agricultural production, human health and migration. The World Bank (2007) is heavily committed to projects that address the short-term impacts of climate variability and to reducing climate change vulnerability. Through the Kyoto initiatives, developing countries may also benefit from the Clean Development Mechanism which could play a significant role in encouraging sustainable investments.

5 Conclusion

In this chapter, we have described an explosive interaction between population growth and a world of scarce resources including arable land, food, water and energy resources. We are at the heart of the 'equation of Johannesburg': more energy for economic development and reduced emissions of greenhouse gases. These interactions create political and social tensions globally and, within a number of countries, they threaten political stability. The question of deforestation is a dramatic illustration. Who is going to pay for limiting deforestation? This could be the moment for proposing a New Marshall Plan for the South, a plan associated with the post-Kyoto instruments. A New Marshall Plan is needed because the situation is dramatic and because the current instruments and mechanisms, both in the public and private sectors, are not sufficient to provide sufficiently rapidly a dynamic of sustainable development. The objectives of the Marshall Plan were 'to fight against hunger, poverty, desperation and chaos'. Do we not face the same problems today? Poverty is a threat to peace and freedom.

References

Alesina, A., Devleeschauwer, A., Easterly, W., Kurlat, S. and Wacziarg, R. (2003) 'Fractionalization', *Journal of Economic Growth*, 8: 155–94.

Aoun, M. C. (2008) 'La rente pétrolière et le développement économique des pays exportateurs'. PhD dissertation, Université Paris-Dauphine.

Bacon, R. and Kojima, M. (2006) *Coping with Higher Oil Prices*. UNDP/ESMAP.

Baumol, W. J. and Lee, K. S. (1991) 'Contestable Markets, Trade and Development', *The World Bank Research Observer*, Washington.

Birol, F. (2007) 'Energy Economics: a Place for Energy Poverty in the Agenda?' *The Energy Journal*, 28, 3: 1–6.

Broadman, H. G. (2008) 'China and India Go to Africa: New Deals in the Developing World', *Foreign Affairs*, 87, 2: 95–109.

Butin, V. and de Gromard, C. (2006) 'L'expérience marocaine d'électrification rurale globale', in S. Michaïlof (ed.), *A quoi sert d'aider le Sud?* Paris: Economica.

Curtain, R. (2002) 'Promoting Youth Employment through Information and Communication Technologies (ICTs): Best Practices Examples in Asia and the Pacific, 2001', paper presented at the ILO/Japan Tripartite Regional Meeting on Youth Employment in Asia and the Pacific, Bangkok, 27 February–1 March.

Hammond, A. L. (2001) 'Digitally Empowered Development', *Foreign Affairs*, 80, 2: 96.

IBRD/World Bank (2008) *MDGs and the Environment. Agenda for Inclusive and Sustainable Development.* Global Monitoring Report. www.worldbank.org

International Energy Agency (2002) *World Energy Outlook*, chapter on 'Energy and Poverty'.

International Telecommunication Union (ITU) (1999) 'Telemedicine in Gambia: Internet for development, challenges to the network'. www.itu.int

Kaufmann, D., Kraay, A. and Mastruzzi, M. (2007) 'Governance Matters VI: Governance Indicators for 1996–2006', *World Bank Policy Research Working Paper* 4280.

Joubert, M. (2006a) 'Favoriser l'accès aux services financiers: microfinance, de l'expérimentation à la professionnalisation du secteur, évolution des pratiques et des idées', in S. Michaïlof (ed.), *A quoi sert d'aider le Sud?* Paris: Economica, 399–413.

Joubert, M. (2006b) 'Savoir faire les bons choix pour créer les bases du succès: AMRET, institution de microfinance exemplaire en milieu rural au Cambodge', in S. Michaïlof (ed.), *A quoi sert d'aider le Sud?* Paris: Economica, 415–32.

Keppler, J. H. (2007) 'Energy Intensity and Prices: Panel Data Analysis', in J. H. Keppler, R. Bourbonnais and J. Girod (eds), *The Econometrics of Energy Systems*. Basingstoke: Palgrave Macmillan.

Michel, S. and Beuret, M. (2008) *La Chinafrique*. Paris: Grasset.

Reddy, A. K. N. (2000) 'Energy and Social Issues', in J. Goldemberg (ed.), *World Energy Assessment: Energy and the Challenge of Sustainability*. New York: UNDP, 39–60.

Sagar, A. D. (2005) 'Alleviating Energy Poverty for the World's Poor', *Energy Policy*, 33, 11: 1367–72.

Stern, N. (2006) *The Economics of Climate Change* (Stern Report). Cambridge: Cambridge University Press.

UNDP Annual Report 2007.

UNDP/ESMAP (2005) *The Impact of Higher Oil Prices on Low Income Countries and on the Poor*. UNDP.

Vincent, K. (2004) *Creating an Index of Social Vulnerability to Climate Change for Africa*. Norwich, UK: Tyndall Centre for Climate Change Research, Working Paper 56.

World Bank (2007) *An Investment Framework for Clean Energy and Development*. Washington, DC: World Bank.

World Bank (2008a) *World Development Indicators, 2008*. Washington, DC: World Bank.

World Bank (2008b) 'Financing Energy Efficiency. Lessons from Brazil, China, India and Beyond', by R. P. Taylor, C. Govindarajalu, J. Levin, A. S. Meyer and William A. Ward. Washington, DC: World Bank

World Bank (2008c) *The Welfare Impact of Rural Electrification: a Reassessment of Costs and Benefits*. Washington, DC: World Bank.

World Bank (2008d) 'Towards a Strategic Framework on Climate Change and Development for the World Bank Group'. Washington, DC: World Bank.

Yunus, M. (2008) *Creating a World Without Poverty: Social Business and the Future of Capitalism*. New York: Public Affairs.

5

Oil and Gas Resources of the Middle East and North Africa: a Curse or a Blessing?

Marie-Claire Aoun

The area covering the Middle East and North Africa (MENA) occupies a key position in the geopolitics of energy. The area, which represents 5 per cent of the world's population, contains 66 per cent of world oil reserves and 45 per cent of world gas reserves. Some of these countries are rich or very rich. However, this windfall wealth is unevenly distributed and does not automatically lead to economic development. Actually, many of these countries suffer from what the economists call the 'resource curse' (more specifically here the oil curse). The oil curse creates economic distortions that impede economic development. In addition, oil dependence has a negative impact on the quality of institutions, in particular when it concerns democracy and corruption. For most of these countries, climate change is not considered as a real issue and energy prices are heavily subsidised.

1 The energy wealth of MENA countries

Since oil was discovered in the beginning of the twentieth century, the Middle East has acquired a strategic importance for international superpowers. Among all MENA countries (twenty-one countries according to the World Bank),[1] thirteen are net oil exporters and eight possess vast oil and gas resources (Table 5.1). In 2006, the oil reserves of Iran, Iraq, Kuwait, Qatar, Saudi Arabia, the United Arab Emirates, Algeria and Libya were estimated at 785 billion barrels, 65 per cent of the world oil reserves for 3 per cent of the world's population and 2 per cent of the world's GDP.[2] These countries' oil resources represent about 80 per cent of OPEC reserves. Their gas reserves reached 77 trillion cubic metres, 43 per cent of world gas reserves. These eight countries all figure among the fifteen most important oil-exporting countries in the world. In 2006, they

Table 5.1 Oil and gas resources in the MENA region in 2006

	Net oil exports in million barrels per day	*Proven oil reserves in billions of barrels*	*Oil production in millions of barrels per day*	*Gas reserves in trillion cubic metres*	*Gas production in billion cubic metres*	*Population 2006 in millions*
Saudi Arabia	8.5	264.3	10.9	7.1	73.7	23.7
Iran	2.5	137.5	4.3	28.1	105.0	70.1
Iraq	1.4	115.0	2.0	3.2		24.4
Kuwait	0.2	101.5	2.7	1.8	12.9	2.6
UAE	2.6	97.8	3.0	6.1	47.4	4.2
Libya	0.2	41.5	1.8	1.3	14.8	6.0
Qatar	1.0	15.2	1.1	25.4	49.5	0.8
Algeria	1.8	12.3	2.0	4.5	84.5	33.4
8 MENA exporters	**18.3**	**785.0**	**27.8**	**77.4**	**387.8**	**165.2**
Oman	0.7	5.6	0.7	1.0	25.1	2.5
Egypt	0.01	3.7	0.7	1.9	44.8	74.2
Syria	0.2	3.0	0.4	0.3	5.5	19.4
Yemen	0.2	2.9	0.4	0.5	–	21.7
Bahrain	0.02	0.1	0.05	0.1	11.1	0.7
Total MENA oil exporters	**19.4**	**800.2**	**30.1**	**81.2**	**474.3**	**283.8**
Share in the world (in %)	**37**	**66**	**37**	**45**	**17**	**4**

Iraq: data for population 1999.
Source: BP Statistical Review of World Energy 2008, Energy Information Administration, World Development Indicators 2007.

exported more than 18 million barrels per day. Access to these very cheap resources remains vital for the functioning of the world economy.

Despite geographical diversification efforts, the world oil dependence on the Middle East is still large and will keep on growing in the coming decades, according to the projections of the International Energy Agency. The Gulf Arabic countries, which belong to the Gulf Cooperation Council[3] (GCC), remain the main suppliers of crude oil to the world market.

MENA oil and gas exporting countries can be divided in two categories. On the one hand, Gulf monarchies like Kuwait, Qatar and the United Arab Emirates are characterised by small populations, vast hydrocarbon resources, high revenues per capita and very low water resources. They are also considered as labour-importing economies, as they have a severe shortage of manpower. On the other hand, some countries like Algeria, Iran, Iraq and Libya have large populations and lower oil reserves per capita. For some of these countries, oil rents have been related to the international context (sanctions). A key country lies in the middle: Saudi Arabia, with a huge resource endowment and a large and fast-growing population (Chevalier and Aoun 2007).

With a population of about 24 million inhabitants, Saudi Arabia has the largest proven oil reserves of the world. The size of the fields and their flexibility enabled the kingdom to play an important role for many years as a market regulator, a swing producer, by modulating its production between 8 and 11 million barrels per day. Its capacity surplus enabled the kingdom to change the direction of prices during the 1990s. The fiscal situation of Saudi Arabia has a major influence on the evolution of crude prices. Due to its spare capacity (which has been drastically reduced since 2004), any decision to increase or decrease its production immediately prompts a market reaction. This production flexibility provides considerable power to the kingdom on the geostrategic world scene.

The dramatic rise of oil prices has led to massive revenue transfers towards oil-exporting countries since 2004. The net oil export revenues of MENA countries reached soaring levels during the last few years. For the year 2007, OPEC net oil export revenues attained $673 billion in nominal terms, of which $524 billion went to MENA countries ($194 billion for Saudi Arabia, $63 billion for the United Arab Emirates, $57 billion for Iran and $55 billion for Kuwait).[4] These levels of revenues were the highest ever earned by these governments. In real terms, in 1980, OPEC oil export revenues were $535 billion (in constant 2000 dollars) in 1980, $554 billion in 2007.[5]

These large revenue streams have enabled many oil-exporting countries, especially GCC countries, to be the major lenders in the international capital market. In 2006, the current account surplus of the GCC member states was $205 billion (for 35 million inhabitants). It was almost equivalent to that of China ($250 billion for 1,300 million inhabitants), whereas the US current account deficit was almost equivalent to $790 billion. These countries' surpluses are now major contributors to global adjustment (Setser 2007).

Moreover, at the end of 2007, GCC countries possessed a considerable investment capacity, estimated at about $100 billion. About 57 per cent of these assets were invested in Europe, 25 per cent in North America and 14 per cent in Asia.[6]

2 Oil rents and the world surplus

Oil-exporting countries are often considered as rentier states, as their economies are highly dependent on external flows. Before analysing the impact of these windfall profits on oil economies, we develop, in this section, the theory that lies behind the economic rents and then analyse the various rents specific to the petroleum industry.

2.1 Theoretical background

Rent theory is generally attributed to David Ricardo, who considered rent as a gift, due to the scarcity and the differential quality of land. Land rent is founded on the heterogeneity of cultivated lands; it is 'paid to the landlord for the use of the original and indestructible powers of the soil' (Ricardo 1821: 53). Rent has no productive counterpart and is only founded on property rights that limit the access of capital to land and mineral deposits (Yashir 1988). The cultivation of less fertile land with higher production costs (due to the increase in the demand for food) creates a differential rent for the more fertile land.

In an article published in 1931, Hotelling demonstrated that in the mining sector, in addition to their production costs, firms have to support an opportunity cost. This cost represents the net present value of the marginal profit at each moment over the extraction period, for an optimal extraction of the non-renewable resource. Thus, a firm will decide to produce the resource only if the market price is high enough to cover the opportunity cost (the net present value) (Otto et al. 2006). The firm faces an arbitrage between the extraction and the selling of the resource today, against the future loss of revenue that could have been earned if the resource was not extracted. This opportunity cost is often referred to as Hotelling rent or scarcity rent. The resource market price thus has to be higher than its marginal cost and includes a scarcity rent. Hotelling shows that there is an optimal path of resource depletion associated with an optimal evolution rate of the resource price. The net price of a non-renewable resource has to increase with the interest rate (the discount rate) to ensure an optimal exploitation of the field. If the price increases faster than the interest rate, the producer would be better off if the resource extraction were postponed, as it represents a more attractive investment than what is offered in the financial market. The Hotelling rule is confirmed if the producer has difficulty in choosing between extracting and transforming the production into a financial asset, or preserving it in the field.

2.2 The rent in the oil industry

The oil industry generates two types of rents: differential and monopoly rents earned by several actors along the oil chain. Differential rents appear with the diversity of production costs in this industry. This type of rent provides an additional profit to the producer whose individual cost is lower than the general production price in the industry. The crude production price is determined by the producers that have the highest individual costs, which generates a differential rent for the other producers.

Mining rent represents the most important share of the differential rents. It is defined as the difference between the extraction costs in two different fields for a given consumption market.

In addition to mining rents, one must consider other characteristics of the natural resource that generate rent differentials. For any given crude, three types of rents can be distinguished (Chevalier 1973):

- A quality rent, due to the chemical composition of the extracted crude (its density, API gravity and its sulphur content).
- A position rent, due to the transportation cost. The proximity of the consumption market confers a position rent for the oil field.
- A technological rent, due to the heterogeneity of the production system on the transportation, refining and distribution levels.

Thus, low production costs in the oil industry generate enormous rents. Figure 5.1 shows the world rent of some primary commodities, following data given by the World Bank.

Oil prices amounted to $41 in 2004 and $65 in 2006 (in constant 2006 dollars), that is a 60 per cent increase. This rise had a direct impact on the

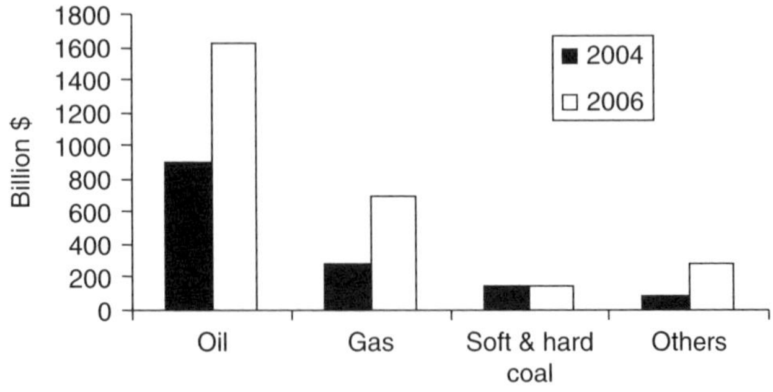

Figure 5.1 World natural resources rent in 2004 and 2006
Rent = (production volume) (international market price – average unit production cost)
* Others: copper, bauxite, lead, phosphate, nickel, zinc, gold and iron
Source: Author's calculation, based on the World Bank Genuine Savings Database (Bolt et al. 2002).

amount of rents generated by the industry. The oil mining rent reached $902 billion in 2004 and $1,628 billion in 2006. These amounts are far beyond gas and coal rents which reached $696 billion and $152 billion respectively in 2006. The aggregation of all the rents from other resources attained $275 billion. This impressive rent differential is due to the low production cost for oil relative to the general price, and to the important difference between the lowest and the highest production costs. Cost diversity generates a differential rent for the companies that explore the favourable fields. In some oil fields in the Middle East, mining rents often represent three to five times the production cost. With respect to similar gas fields, rents vary from 33 per cent to 100 per cent of production costs.

The oil industry also enjoys a monopoly rent which appears when the availability of a commodity is artificially restricted. It covers the excess profit rate realised in the petroleum industry relative to the profit rate in other industries. Two factors create monopoly rents in the oil market: entry barriers and the non-substitutability characteristics of some petroleum products.

Monopoly rents are first earned by producing countries, more specifically OPEC members, as they have the lowest production costs. These countries enjoy a market power, due to their spare capacity, that enables them to capture a monopoly rent by artificially maintaining the price above a floor price. After the oil price collapse in 1998, OPEC members, especially Saudi Arabia, decided to reduce their production and maintain crude prices in a range of $22 to $28. This price band resisted the geopolitical events of 2003 and the production cut that resulted from the political crisis in Venezuela, social unrest in Nigeria and war in Iraq. During this year, Saudi Arabia, with Kuwait and the Emirates were able to compensate for the 'missing barrels'. The Saudi production soared from 9 million to 10.2 million barrels per day between 2002 and 2003. Since 2004, crude spare production capacity has been relatively low (between 1 million and 1.5 million barrels per day).

It is important to highlight that the oil industry does not follow Ricardian logic. Low cost producers (OPEC members) cut down their production to maintain prices at a higher level than in a competitive market.

Another monopoly rent is captured by consuming countries, through domestic taxation on petroleum products (gasoline, diesel oil). The size of the oil rent depends on the price elasticity of oil product demand and on the possibilities of substitution (biofuel). These taxes represent more than 50 per cent of the composite barrel price in some European countries.

As explained in Chapter 1, the sum of all these rents (differential and monopoly rents) represents the world oil surplus. It is the difference between total sales of petroleum products in the world and the total costs incurred for discovering, extracting, transporting, refining and distributing these petroleum products. According to our calculation,[7] in 2004, global oil sales amounted to about $2,525 billion, with corresponding costs estimated at about $545 billion (production, transportation, refining and distribution costs). Therefore, the total oil surplus reached about $1,980 billion, with 59 per cent earned by consuming countries through taxation on petroleum products, 35 per cent by governments of producing countries due to their fiscal instruments during the production process and 7 per cent by all corporate players present in the chain (through their industry margin).

The analysis of this 'pie' sharing reveals that oil is an extremely political energy. About 93 per cent of the surplus generated by the production and trade of petroleum products is mainly controlled by governments rather than by private actors.

In this setting, oil receipts confer substantial financial power and a strategic position on the international scene. However, these streams are also a source of vulnerability, as oil-exporting economies remain completely dependent on these resources and are subject to oil price volatility.

3 The resource curse

The evolution of oil prices between 2004 and 2008 generated high economic growth rates in the region. However, the economic performances of oil- and gas-rich MENA countries have been particularly disappointing during the last three decades, despite their huge hydrocarbon resources. This pattern of underdevelopment reveals a counter-intuitive phenomenon, widely observed in resource-rich countries. Natural resources, above all, oil resources, impose a limit on growth opportunities. The development experiences of many countries prove that the possession of natural resources is often transformed into a curse. Moreover, in some cases, oil wealth has exacerbated civil conflicts due to the permanent struggle for the appropriation of the rents.

3.1 Economic performances of MENA oil-rich countries

When we examine the growth rates of world net oil exporters, we perceive the poor performance of their economies (excluding Norway) during the last three decades. The average economic growth rates of emerging 'resource-poor' countries are much higher. Alas, the GDP per capita of major net oil-exporting countries has been declining since 1975. Between 1975 and 2005, the world average GDP per capita[8] increased by 75 per cent, from $4,834 to $8,476. During the same period, the average GDP per capita for oil-rich MENA countries declined by 23 per cent, whereas that of oil-poor countries in the same region (Jordan, Tunisia and Morocco) nearly doubled (Table 5.2).

This decline can be partially attributed to high demographic growth rates in the region. Between 1975 and 2005, the average population growth rate was 2.1 per cent in MENA oil-rich countries, well above the world average (1.2 per cent). All Middle East countries must now cope with high population growth rates (among the highest in the world). The average population growth rates between 1985 and 2006 are 5.4 per cent for the United Arab Emirates and 3.6 per cent for Kuwait.[9] The population growth of Saudi Arabia has far outpaced the growth of its economy, and the level of oil reserves relative to population dropped from 16,000 barrels in 1983 to 11,000 in 2006. Disappointing growth performances reveal the failure of many oil-rich governments to promote long-term economic policies that support these dynamic demographic trends.

Concerning social indicators, the Human Development Index shows that development levels are relatively high in GCC countries, due to the cohesive system adopted by governments. However, the comparison of development performance with economic prosperity reveals significant shortages in social policies concerning the sectors of health and education (components of the HDI).[10] In order to evaluate the efficiency of social policies, the United Nations measures the difference between the

Table 5.2 Economic development and GDP per capita of MENA oil-rich countries

	GDP per capita, PPP (constant 2000 international $)*			*Human Development Index (HDI) 2005*		*Population 2006 Growth rate %*
	1975	*2005*	*Average annual economic growth 1975–2003*	*Rank (177 countries)*	*Index*	
Algeria	4,712	6,361	0.4%	104	0.733	1.5
Libya	3,732	7,517	1.0%	56	0.818	2.0
Bahrain	11,479	14,588	0.4%	41	0.866	1.9
Egypt	1,504	3,985	1.4%	112	0.708	1.8
Kuwait	27,760	19,791	–0.5%	33	0.891	2.5
Oman	4,504	8,961	1.2%	58	0.814	1.6
Saudi Arabia	21,590	13,175	–0.7%	61	0.812	2.4
UAE	44,601	22,109	–1.1%	39	0.868	3.5
Iran	6,985	7,137	0.0%	94	0.759	1.5
Syria	2,404	3,437	0.5%	108	0.724	2.7
MENA exporting countries average	**14,681**	**11,252**	**–0.4%**		**0.799**	**2.1**
MENA importing countries average **	**2,753**	**5,321**	**1.0%**		**0.728**	**1.5**
World average	**4,834**	**8,476**	**0.8%**			**1.2**

* PPP: Purchasing Power Parity
** Average for Jordan, Morocco and Tunisia
Source: World Development Indicators 2007; Human Development Report 2007/2008.

country's wealth and the actual development of its human resources (since countries with similar income can have very different HDI values). Most MENA oil-rich countries have large negative values for GDP per capita (PPP$) rank minus HDI rank (Algeria (−22); Iran (−23); Saudi Arabia (−19); Qatar and the United Arab Emirates (−12) and Oman (−15)).[11] These negative results suggest that these countries have failed to translate their economic prosperity into correspondingly better lives for their people and positive social development. On the other hand, some oil-poor MENA economies, like Jordan (+22), for example, realise a positive figure, which shows that income was converted into economic development very effectively.

The favourable conjuncture of the oil market between 2004 and 2008 generated high growth rates for MENA oil-rich countries. The average real GDP growth rate reached 6.4 per cent for oil exporters in the MENA region and 6 per cent for GCC countries in 2006 (6.3 per cent for Kuwait, 10.3 per cent for Qatar, 9.4 per cent for the United Arab Emirates and 4.3 per cent for Saudi Arabia) (IMF 2008). These growth performances were among the best in more than ten years.

Despite these recent growth trends and their substantial financial power on the international scene, MENA oil economies still have to deal with several challenges. The Saudi economy, for instance, needs a rate of economic growth of 6 per cent in order to support a growing young population (more than half of the population is less than 25 years old). According to Cordesman (2006), the rate of unemployment is estimated at about 8–13 per cent for Saudi males in 2004–2005, 16.6 per cent for males aged between 20 and 29 years old. The maintaining of a high fertility rate (5.3 births per woman in Saudi Arabia) and the population increase will certainly have dramatic implications on its labour market and educational system, as well as on the size of future affordable subsidies. In Bahrain, unemployment is estimated at between 12 per cent and 18 per cent of the workforce, due to the entry of many youths into the labour market.

3.2 The resource curse: the 'Dutch Disease' and lack of governance (the rentier states)

The negative impact of economic dependence on primary exports, that is, the resource curse, has been widely analysed by economists. Stevens (2003) offers a survey of the studies that established a negative relation between natural resources (mining, agriculture and hydrocarbons) and economic growth.

3.2.1 The 'Dutch Disease'

The theoretical background underlying the resource curse is roughly divided into two major trends. The 1970s shocks incited researchers to explore the impact of oil revenues on the economy. In the 1980s, the 'Dutch Disease' was the major explanation of the phenomenon (Corden and Neary 1982). This expression appeared for the first time in an article in *The Economist* (1977) which described the disappointing economic experience of the Netherlands after the exploitation of the Groningen natural gas reserves. This term refers to an odd contrast between, on the one hand, a morose economic situation (low industrial production, decreasing investments and profits and rise of unemployment) and on the other hand, an external balance surplus (strong guilder, high surplus in the current account). In fact, the real exchange rate appreciation entails a contraction, if not the destruction, of non-oil tradable goods. This sector becomes progressively uncompetitive. One then observes a massive labour force transfer towards the oil sector, because production factors are better remunerated there. Hence, oil-exporting countries gradually become largely dependent on oil revenues. Furthermore, positive externalities inherent in the manufacturing sector progressively disappear: this sector generates 'learning by doing', the acquisition of know-how and technological progress (Torvik 2001).

3.3 Oil and gas dependence

The structure of the trade balance of oil-exporting countries indicates that oil sales represent a very high share of export revenues. These countries are not able to export anything but hydrocarbons. The rest of the economy (agriculture, manufacturing and industry) is completely distorted by oil and gas wealth, which inflates domestic prices and crowds out any other exports. In the OPEC countries (Indonesia excluded), hydrocarbon exports contribute to about 50 per cent of their total exports and more than 52 per cent of their fiscal resources (IMF 2007). Figure 5.2 shows the degree of dependence of MENA economies on the hydrocarbon sector, between 2000 and 2005.

This over-reliance on petroleum revenues is a direct outcome of the Dutch Disease phenomenon. The extractive industry is a highly capital-intensive and enclave-oriented industry, that is, there is a lack of productive linkages with the rest of the economy. For example, whereas the hydrocarbon sector contributes more than 36 per cent of the GDP (in value added) and 97 per cent of total exports in Algeria, it hardly employs 2 per cent of the population. Conversely, agriculture and manufacturing contribute together 30 per cent of GDP and 42 per cent of employment.[12]

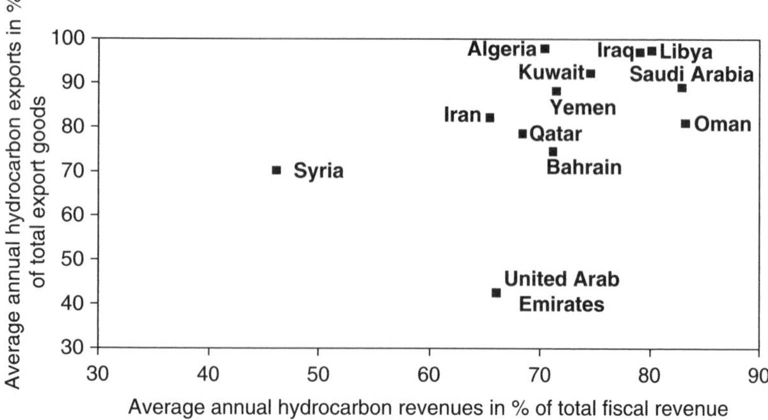

Figure 5.2 Dependence of MENA economies on the hydrocarbons sector (average 2000–2005)
Source: IMF (2007).

The unique connection between the oil sector and the rest of the economy is realised through fiscal resources generated by oil production that feed the state budget. This extends the role of the state and weakens that of the private sector. Hence, these countries rapidly reach the limits of their economic absorption capacity (Askari and Jaber 1999).

During the last two decades, governments of the GCC countries have been using fiscal policy as a primary instrument to achieve economic objectives such as promoting economic growth and reducing unemployment and revenue inequalities. The monetary policy was essentially directed at stabilising the exchange rate and controlling inflation (in order to avoid the adverse effects of the Dutch Disease). Several constraints have been diminishing the flexibility of the fiscal policy: a heavy dependence on volatile oil export revenues, a very important wages bill (civil servants' salaries), various subsidies and, in the case of Saudi Arabia, high domestic debt service until 2004 (Fasano and Iqbal 2003).

Concerning monetary policy, the GCC countries formally pegged their currencies to the dollar in January 2003, as a first step towards a monetary union, which is scheduled for 2010. However, the continuous depreciation of the dollar since 2004 is generating a considerable loss of purchasing power, as most MENA countries trade increasingly with the euro and Asian zones, and less with the dollar zone. The dollar-peg also feeds inflation which is a major concern for all GCC countries. Indeed, inflationary pressures have emerged with the oil boom, with an average consumer price inflation of 6 per cent for GCC countries in 2007. In May 2007, Kuwait decided to abandon the peg, concerned that the weakening of the US currency was fuelling domestic inflation. Since that decision, the debate over the long-term viability of the GCC's currency peg has been growing (IMF 2008). Furthermore, the adoption of a more flexible exchange rate seems to be necessary in order to increase competitiveness of non-oil exports.

3.4 Poor governance and rentier states

Good governance and efficient institutions have emerged progressively in the 1990s as the core element of any development strategy. Many researchers explained the resource curse by the adverse effects of primary resources on a country's institutional quality. During the last decade, the resource curse has been progressively tied to the political economy: natural resources tend to impede the development of sound economic and political institutions.

For example, Karl (1999) explains why the high primacy of the state in oil-producing countries leads to 'petromania' behaviour and to rent addiction. Political leaders, rulers and the economy's elite adopt rent-seeking activities, which reduce incentives for entrepreneurship and cause individuals to drift away from productive and innovating activities. Furthermore, as oil rents are easily appropriated by the state – in the cases of Iran, Libya and Saudi Arabia – governments are relieved of the pressure of taxation. The virtuous circle of creating wealth by the protection of property rights and the accountability of government disappears. Thus, there is no incentive to develop growth-friendly political institutions (Birdsall and Subramanian 2004).

Besides, rent-seeking activities often give rise to corruption. Leite and Weidmann (1999) demonstrated that capital-intensive natural resource industries are a major cause of corruption. Corruption is a permanent feature in oil and gas activities in many developing countries. A significant part of hydrocarbon rents is diverted from official flows and often goes directly to individuals or groups in power positions. About $12 billion of natural resource revenues were grabbed worldwide by rebels, corrupt governments and predatory groups in the 1990s (O'Higgins 2006).

The opacity index calculated by the Kurtzman Group measures the cost to businesses of a lack of transparency in a country's legal, economic, regulatory and governance structures. Investment decisions of international companies are closely related to the perception of 'small-scale risks', such as fraudulent transactions, bribery, regulatory complexity and unenforceable contracts. These risks have a direct impact on businesses as they deter investment. The index considers five broad causal categories (which form the acronym CLEAR): business and government Corruption, an ineffective Legal system, deleterious Economic policy, inadequate Accounting and governance practices, and detrimental Regulatory structures. Each score is associated with an opacity risk premium (or discount) expressed as an interest rate equivalent. This index suggests that doing business in Saudi Arabia, which has an opacity score of 46 (out of 100), requires a return of 5.52 per cent above the US rate of return in order to offset a company's risk. In the same manner, the lack of transparency in Egypt requires a premium of 5.91 per cent (opacity score 48) (Kurtzman et al. 2004).

Natural resources also affect the political regime. Ross (2001) has shown an unexpected relation between oil wealth and political outcomes. His econometric study for 113 states between 1971 and 1997 revealed that oil resources tend to inhibit democracy. Heavily oil-dependent states would lose 1.5 points on the democracy scale due to their oil wealth alone. This negative impact is more important in poor states than in rich states. To explain the link between authoritarianism and oil wealth, Ross (2001) distinguished three complementary effects: a 'rentier' effect, government's tendency to use low tax rates and high spending in order to appease pressures for democracy and preventing the formation of social groups independent of the state. Without the pressure of taxation, there is no incentive to create mechanisms of accountability and to develop civil society. A repression effect such as oil wealth enables authoritarian governments to invest heavily in internal security forces and to block the population's democratic aspirations. Finally, there is a modernisation effect: that is the economic dependence on the primary sector which prevents the population from attaining industrial and service sector jobs. The claim for democracy is therefore weakened. Table 5.3 displays the main governance indicators for the chief oil-exporting countries in the MENA region.

These indicators reveal that there is a significant deficit in public governance in many oil-rich countries. Most countries suffer from weak political and economic institutions. The Transparency International (TI) Corruption Perception Index for the year 2007 shows that oil wealth is a breeding ground for corruption. It places major oil exporters near the bottom of the list with the following rankings: Saudi Arabia at the 79th position, Algeria at the 99th, Iran and Libya 131st and Syria 138th, in a survey that included 180 countries (the 180th being the most corrupt). Moreover, many of these countries are considered by Freedom House as politically 'non-free' states. Most of them have repressive regimes

Table 5.3 Governance indicators in MENA oil-exporting countries

	Transparency International Corruption Perception Index 2007		*Freedom House*			*World Bank governance indicators 2006*					
	Index	*Ranking*	*PR*	*CL*		*VA*	*PV*	*GE*	*RQ*	*RL*	*CC*
Saudi Arabia	3.4	79	7	6	NF	−1.42	−0.65	−0.28	−0.02	0.17	0.18
Iran	2.5	131	6	6	NF	−1.33	−1.25	−0.80	−1.51	−0.81	−0.59
Iraq	1.5	178	6	5	NF	−1.54	−2.91	−1.70	−1.46	−1.95	−1.40
Kuwait	4.3	60	4	5	PF	−0.36	0.28	0.28	0.51	0.75	0.67
UAE	5.7	34	6	6	NF	−0.78	0.68	0.78	0.80	0.67	1.16
Libya	2.5	131	7	7	NF	−1.90	0.24	−0.86	−1.40	−0.74	−0.89
Qatar	6	32	6	5	NF	−0.51	0.86	0.53	0.45	0.93	0.83
Algeria	3	99	6	5	NF	−0.83	−0.89	−0.35	−0.61	−0.63	−0.39
Oman	4.7	53	6	5	NF	−0.77	0.66	0.46	0.75	0.71	0.71
Egypt	2.9	105	6	5	NF	−1.08	−0.87	−0.41	−0.44	0.00	−0.41
Syria	2.4	138	7	7	NF	−1.64	−0.88	−1.03	−1.24	−0.55	−0.66
Yemen	2.5	131	5	5	PF	−1.06	−1.40	−0.93	−0.68	−0.98	−0.60
Bahrain	5	46	5	5	PF	−0.71	−0.42	0.35	0.72	0.62	0.58

Transparency International Corruption Perceptions Index 2007
A country or territory's CPI score indicates the degree of public sector corruption as perceived by business people and country analysts, and ranges between 10 (highly clean) and 0 (highly corrupt). http://www.transparency.org/policy_research/surveys_indices/cpi/2007

Freedom in the World Country ratings: Freedom House
Political rights (PR) and civil liberties (CL) are measured on a 1–7 scale, with 1 representing the highest degree of freedom and 7 the lowest. The resulting status is 'F', 'PF' and 'NF', respectively, standing for 'free', 'partly free' and 'not free'. http://www.freedomhouse.org

Worldwide Governance Research Indicators Dataset: World Bank[13]
The six dimensions of Governance are: Voice and Accountability; Political Stability and Absence of Violence; Government Effectiveness; Regulatory Quality; Rule of Law; and Control of Corruption.
They are reported by Point Estimate: from −2.5 (worst governance) to +2.5 (best governance) (See Chapter 4)

- PV: Political Stability and Absence of Violence
- VA: Voice and Accountability
- CC: Control of Corruption
- GE: Government Effectiveness
- RQ: Regulatory Quality
- RL: Rule of Law

www.worldbank.org/wbi/governance

with a high degree of authoritarianism. The indicators of governance of the World Bank confirm the limits to political freedom. The voice and accountability indicator also reveals that freedom of expression, of association and of the media are particularly restricted in all selected countries.

Sometimes, natural resources, especially oil, have generated situations of extreme institutional collapse. Oil has been frequently associated with civil conflicts and wars, not only in the Middle East, but also in Nigeria with the Biafra rebellion and with the struggle over Cabinda in Angola. Collier and Hoeffler (2004) show that the risk of civil wars is closely related to high levels of dependence on petroleum exports. The Middle East has long been at the centre of struggles for the control of resources and it has been subject to many intra-regional and internal conflicts, as well as religious and ethnic divisions. It is one of the most militarised regions in the world and many conflicts are often motivated by oil access. Military and security expenditures are very high and constitute a heavy burden on these economies. In 2005, military expenditures accounted for 5 per cent of the Kuwaiti and Iranian GDPs and 8 per cent of that of Saudi Arabia (the world average is 2.5 per cent). The Iraqi economy, for example, was structurally modified during the 1980s due to the Iran–Iraq war. The military sector used to employ about 3 per cent of the population in 1975, but it absorbed 21 per cent in 1988 at the end of the war. During the period 1981–8, military spending totalled $120 billion, that is, 256 per cent of the same period's oil revenue of $46.7 billion (Alnasrawi 2002). Undeniably, these expenditures have a negative impact on economic growth. They reduce the domestic investment potential and block the development of the productive economy.

3.5 Redistribution of the oil wealth

Most GCC governments are considered by Eifert et al. (2002) as 'paternalistic autocracies', because they initially based their legitimacy on traditional and religious authority and maintained it through the 'mobilisation of oil wealth to prop up public living standards'. Some observers have identified a 'Santa Claus' effect in rentier states like Saudi Arabia and Kuwait. The government becomes very generous and provides welfare goods to the population to maintain social and political peace. Subsidies and other transfers represent respectively about 12 per cent, 15 per cent and 30 per cent of the public expenditures of the United Arab Emirates, Bahrain and Kuwait. The substantial subsidies may affect the prices of current transactions (for wheat, electricity, hydrocarbons, water desalination) and capital transactions (loans at preferential rates).

Heavy price subsidies lead to inefficient energy use and represent a heavy burden on public budgets. In many Gulf countries, *per capita* energy consumption is among the highest in the world. In 2005, Qatar has the highest energy use, about 20 toe per capita, with 11 toe per capita for the United Arab Emirates and Kuwait, compared to 8 in the United States, 4 in Western Europe and 1.3 in China.[14] In all MENA countries, prices of oil products are held well below international market levels. In GCC states, electricity is priced at a fraction of its true supply cost, or even given away. There is, however, a widespread consensus among policy-makers to reduce these subsidies, but 'indigenous populations have become accustomed to sharing directly in the national

Box 5.1 Inefficiencies in the energy sector and in the water system in the MENA region

In addition to demographic and unemployment challenges, the MENA region also has to deal with a high level of inefficiency in the energy sector and in the water system. There are substantial losses in the water distribution network and the water supply is also heavily subsidised. Water scarcity is a key concern for several oil-rich countries such as Kuwait, Qatar, the UAE, Libya, Algeria and Saudi Arabia. These countries have among the lowest natural renewable water resources in the world. Furthermore, the region has a very low level of energy efficiency in various fields. For example, the efficiency of air conditioning used in commercial buildings and in households is below the world average, as low electricity prices give no incentive to consumers to use more efficient appliances.

High energy use also has a significant environmental impact. The MENA region GHG emissions account for 5 per cent of the total emissions in the world. Iran, Saudi Arabia and Egypt contribute respectively 27 per cent, 18 per cent and 10 per cent of the total emissions in the region. According to a recent study by ESMAP, an improvement by 10 per cent in energy efficiency in the manufacturing sector in the United Arab Emirates could reduce the total CO_2 emissions by 2.86 per cent (Zhang 2008). Consequently, the governments of the MENA region have recently accorded a high priority to energy efficiency due to several pressing domestic concerns such as urban air pollution, energy security, fiscal cost of energy subsidies and the reduction of GHG emissions.

Source: Zhang (2008); IEA (2005).

wealth generated by hydrocarbons, by paying little or nothing for their use' (IEA 2005: 259).

These governments also invest massively in public sector employment, which accounts for more than 70 per cent of total employment (for nationals) among GCCs.[15] Indeed, in Saudi Arabia, the administration is considered as the 'employer of first resort'. Private jobs are therefore often left to the foreign workers (Auty 2001).

During 'hard times', as in the 1990s, when oil revenues collapsed, the ability of these countries to maintain welfare payments and entitlements was drastically reduced and the need for policy reorientation became a priority. The Saudi government was forced to borrow heavily from domestic creditors and thus accumulated, over the years, $170 billion of domestic public debt, that is, more than 92 per cent of its GDP according to the International Monetary Fund in 2002. This massive debt was a burden that clearly limited the ability of the kingdom to launch economic expansion reforms.

Thus, oil rent redistribution by governments through inflexible expenditure commitments, such as various subsidies, high levels of public employment and overstaffed bureaucracies can push many rentier states towards fiscal crisis and lead to heavy inefficiencies.

4 The dynamics of the rent sharing

Among corrective actions adopted by policy-makers of oil-exporting countries, we have identified two preventive policies in order to 'curb' the pace of the curse. Revenue management funds and export diversification away from hydrocarbons are considered as efficient options to combat the adverse effects of the Dutch Disease and reduce dependence on the oil rent.

4.1 The oil funds

Policy-makers of oil-exporting economies have to deal with two major issues. They have to determine an optimal repartition between the present and the future generation, due to the finite nature of the resource. They also have to adjust government spending and cushion the domestic economy against sharp and unpredictable variations in oil prices and revenue (Fasano 2000). Furthermore, they have to consider social unrest in defining their policies and find an optimal strategy for redistribution of oil rents to the population.

Oil funds receive inflows related to the exploitation of oil resources (or other non-renewable resources). They are generally considered as public sector institutions and are separated from the budget (Davis et al. 2003). They can be classified into three types.

1. Stabilisation funds contribute to reduce the volatility impact of oil revenues on the economy. They are designed to support fiscal discipline and provide better transparency of oil revenues. The fund accumulates resources when the oil price exceeds a pre-announced threshold, and pays out when the price falls below a second predetermined threshold. The stabilisation of government revenues and the reduction of volatility and uncertainty make it possible to avoid any disruption in public investment programmes and to attract new investments. Several countries have introduced stabilisation funds: Norway, Venezuela, the United States (Alaska), Oman, Azerbaijan and Chad.[16]
2. Savings funds are designed to put aside oil revenues for future generations, in order to ensure intergenerational equity. They are regularly fed by a share of oil revenues, which is invested in financial assets in the international capital markets. This type of fund was adopted in Kuwait, Alaska and Norway.
3. Funds for the redistribution of oil revenues to the population were implemented in Alaska (the United States) and in Alberta (Canada). In Alaska, the Permanent Fund Dividend Program was created in 1982. It accumulated $37.8 billion between 1982 and 2007. About 41 per cent of its revenues were redistributed to the population as dividends ($15.7 billion), that is, about $1,059 per year on average for each resident.

The most successful oil fund experience is the Norwegian Government Petroleum Fund (a stabilisation and savings fund) established in 1990. It accumulated more than $340 billion in October 2007 and pursues socially responsible investments. This fund is expected to finance a significant share of the retirement income of the post-war baby-boomers in Norway over the next twenty years.

During the last few years, oil funds have become increasingly popular. In October 2007, among the twenty-two most important sovereign funds in the world, fourteen were financed by oil and gas reserves, that is, 72 per cent of the total amount of assets. Table 5.4 shows the main oil funds created in MENA oil- and gas-rich countries.

The Abu Dhabi Investment Authority (ADIA) is considered to be the largest sovereign wealth fund in the world. It is a public organisation

Table 5.4 Oil and gas sovereign funds in the MENA region (assets in October 2007)

Country	*Name of the fund*	*Assets (in billions of dollars)*	*Date of creation*
United Arab Emirates	Abu Dhabi Investment Authority (ADIA)	875	1976
Saudi Arabia	Various funds	300	NA
Kuwait	Reserve Fund for Future Generation	250	1960
Libya	Oil Reserve Fund	50	2005
Algeria	Fund for the Regulation of Receipts	42	2000
Qatar	Qatar Investment Authority	40	NA
Iran	Oil Stabilisation Fund	15	1999
Oman	State General Stabilisation Fund (SGSF)	8.2	1980
Total MENA oil and gas sovereign funds		**1580**	
Total (22 most important world sovereign funds)		**2827**	

Source: *Le Monde* (2 October 2007), Morgan Stanley data.

managed by the federal government of the United Arab Emirates. ADIA was created in 1976 and its main funding source is oil export revenues. The fund has never disclosed the size of its assets. The IMF estimated the size of its assets at about $875 billion in 2007 (Sénat 2007). Setser and Ziemba (2007) assess that ADIA controls assets equal to about 1 per cent of global market capitalisation. The Kuwait Investment Authority (KIA) has also attained considerable size and is considered as a powerful force in the global market. The combined assets of GCC sovereign funds account for over half of the sovereign wealth funds globally.

However, as suggested by Truman (2007), the management of these funds has become a key concern of international policy, because of their size, their potential to disrupt financial markets and their lack of transparency. Indeed, there is a significant lack of information concerning their asset structure and investment strategy.[17] The state of Kuwait established a law in 1982 which stipulates that 'the disclosure to the public of any information related to KIA's work is subject to penalties'. In the same manner, financial operations of the Qatar Investment Authority are not reported. This opacity can entail biased perceptions of government wealth and distort the analysis of fiscal balance and the solvency of the country.

Iran also established an Oil Stabilisation Fund in 2000 in order to reduce the effect of oil price volatility on the government budget and to promote the private sector. However, the fund has been used to top up budget spending. Since its creation, withdrawals from the fund have been particularly high, to finance a number of expenditures, such as the importation of refining products or to fill in public deficits. The Iranian fund lacks a clear separation between the share that should be used to cushion against price fluctuations and the share that should be kept for long-term savings (IEA 2005).

4.2 Export diversification policies: the case of the United Arab Emirates

Greater economic diversification enables an oil exporter to mitigate the negative effects of oil price volatility. Diversification towards non-oil activities also leads to competition, prompts innovation and attracts investment. Exports of manufactured products and services promote long-term economic growth as these sectors require a qualified workforce and technology. The development of a non-oil sector thus creates more employment opportunities. Only a few oil exporters succeeded in their diversification strategy: Malaysia, Indonesia, Mexico, Norway and the United Arab Emirates.

The United Arab Emirates[18] have been going through one of the most impressive economic transformations of the last few decades. In 1980, about 90 per cent of total exports came from the oil sector. Between 2000 and 2005, non-hydrocarbon exports accounted for 57 per cent of total exports on the average (according to the IMF 2007). The pace of reduction in oil dependency was the fastest among all GCC countries. The petrochemical sector, aluminium, tourism and the warehousing trade experienced a real growth rate of 9 per cent during the 1990s.

The development of the economy was based on a strong open trade regime and unrestricted capital outflows, a well-developed physical and institutional infrastructure and a deregulated and competitive business environment with low taxes. The success of the non-oil sector was also facilitated by a rapid expansion of the services sector in the areas of tourism, finance, transport, port facilities and communication (IMF 2005).

Each of the Emirates has been pursuing an economic strategy according to its comparative advantage. Abu Dhabi is specialised in large-scale capital and energy-intensive downstream industries, Sharjah in small-scale light manufacturing and tourism and Ras al-Khaimah in cement and pharmaceutical products. As for Dubai, this Emirate has been developing a manufacturing industry and has become the leading financial centre in the Middle East and in the world. The size of Dubai's economy nearly doubled in the 1990s, with an annual rate of growth of 16 per cent and a 6 per cent contribution from the oil sector. This economic success was due to various policies such as the creation of free trade zones, a favourable investment climate and a high degree of technology absorption (IMF 2005). The government has also built high tech centres attracting many companies, from Microsoft to IBM.[19]

In this setting, the building and construction sector has become the third largest sector in the Emirates after the oil and trade sector, and is among the largest and fastest-growing construction markets in the world. However, this sector, as well as the entire UAE economy, remains extremely dependent on foreign workers.[20] In 2005, about 22 per cent (600,000) of the migrant workers were employed in the construction sector. These workers mainly came from India, Pakistan, Bangladesh and Sri Lanka. More than half of them were employed in Dubai.

As for the skilled workforce, the government of the UAE has adopted liberal labour policies, which allow the recruitment of mainly expatriate workers at internationally competitive wages. According to the IMF (2005), whereas job creation was more than sufficient to absorb the new entrants to the labour force, unemployment among nationals increased gradually between 1999 and 2004. This is due to the fact that about 90 per cent of nationals are employed in the public sector, which provides generous wages (as well as transfers and subsidies) compared to those of the private sector, greater job security, shorter working hours and safer prospects for promotion. Moreover, the skill levels of a significant share of national college graduates do not meet the high standards of national companies and the needs of the private sector.

In order to promote employment among nationals, the government is taking several measures to increase the cost of expatriate labour and is applying quotas to increase the employment of nationals in the private sector. Measures are also being taken in order to reduce the disparity between public and private sector wages.

Concerning the other MENA economies, the development of their non-oil industries has been more limited. The other GCC countries (except Bahrain) have not yet found non-oil export niches. Their export diversification strategy remains mainly driven by the development of the refining and petrochemicals industries. These industries account for 76 per cent and 38 per cent of the manufacturing sectors in Kuwait and Saudi Arabia respectively.

Box 5.2 Nuclear energy in the Middle East and other oil- and gas-exporting countries

Many oil- and gas-exporting countries have expressed an interest in building nuclear power plants. For example, Egypt announced plans to have a 1 gigawatt nuclear power plant operating in 2018, Indonesia's government planning has allocated funds for 6 GWe of nuclear capacity in operating by 2025 and Iran, which is building one 1 GWe nuclear power plant, has announced plans for two more GWe by 2020.

Indeed, with high oil and gas prices, it could be worthwhile to export more oil and gas, or to save it for future generations, instead of using it to produce electricity. Therefore, these countries are interested in producing at least part of their electricity and, in some cases, desalinating water, by using nuclear energy.

The main difficulty in developing nuclear for these countries is the lack of local infrastructure and expertise. Building from scratch the necessary infrastructure and expertise to operate nuclear facilities and implement the necessary safety rules and regulations can take a long time – up to 12 or 15 years.

An interesting solution is under consideration by the United Arab Emirates, specifically Abu Dhabi. It consists in signing a contract with three French companies – Total, Suez and Areva – for building and operating two EPR nuclear power plants totalling 3.2–3.4 GWe. At least 50 per cent of the financing would come from the UAE. The electricity produced would be sold to the Abu Dhabi Electricity and Water Authority under a long-term power purchase agreement. Under the most optimistic scenario, advanced by French officials, this plant would not produce electricity until 2017, but likely later, since the UAE has to develop the minimum capability of a nuclear investor as well as an independent, competent and efficient nuclear safety authority. In spite of cooperation in place with the French Atomic Energy Commission and reliance on outside (US) consultants, this must take a minimum time, at least five years and probably more, including the licensing process. This must be added to the time needed for construction of the plant, expected to be a minimum of six years. That means a likely minimum lead time of eleven years before the plant can come online. There are also delicate issues to be resolved, notably concerning the back end of the nuclear fuel cycle (spent fuel and waste management).

> This arrangement, if implemented, would be the first in the world where foreign companies will build and operate a nuclear power plant and sell the electricity produced. A similar idea was considered for Turkey over a decade ago but that project never came to fruition.
>
> C. Pierre Zaleski (CGEMP).

In many GCC countries, however, the role of the state is declining relative to the private sector due to trade liberalisation, price and investment reforms and the restructuring or privatisation of many public companies (Al Moneef 2006). In Saudi Arabia, for example, the private sector was structurally reformed in order to ease business start-ups. Until July 2007, the minimum capital requirement for starting up a business in Saudi Arabia was the highest in the world. It amounted to fifteen times the average GDP per capita. These requirements used to hamper the development of the private sector. In 2007, Saudi Arabia simplified business start-up procedures. The time to start a business fell from thirty-nine days in 2006 to only fifteen days in 2007. According to the report 'Doing Business', the kingdom's rank in terms of ease of starting a business soared from 159 in 2007 to 26 in 2008 (Belayachi and Haidar 2008).

The oil industry has many distinctive features. It involves huge amounts of rents and provides a powerful position for exporting countries on the international scene. Before the oil price decrease in 2008, MENA governments built up considerable financial reserves. However, oil wealth is a source of profound fragility as the economy is subject to crude volatility. The oil sector is also particularly exposed to poor governance due to unusual windfalls, its enclave nature and the implication of few individuals in the production system. Except for the United Arab Emirates, these countries remain highly dependent on oil receipts, despite several diversification efforts. The overwhelming public sector as well as the large subsidies impose a heavy burden on the fiscal budget and generate serious inefficiencies. The development of a long-term strategy in order to promote the private sector and generate enough jobs for the growing population remains a priority.

The Middle East and North Africa will play an important but very unclear role in the history of the century. The area is a mosaic of various countries with a strong potential for tension and violence. The question of Palestine, the Iranian nuclear threat, the total disorganisation of Iraq, and the ambitions and strategy of Al Qaeda in the area are major potential sources of violence. Many countries of the area are majority Muslim but Islam does not represent a unified view of the political and economic organisation of the society. Within each country there are also sources of potential conflict due to ethnic fractionalisation, religious oppositions and income inequalities. The combination of poverty, frustration and radical Islam may encourage local and international terrorism. In this chapter, the economic and institutional performances of several countries have been evaluated with typically 'Western' criteria. These criteria are far from being accepted by most governments of the area. This raises a real problem for resolving the new energy crisis: Would a world regulation, taking into account the issues of climate change, be accepted by the majority of the world's population?

Notes

1. The Middle East and the North African regions include Algeria, Bahrain, Djibouti, Egypt, Iran, Iraq, Israel, Jordan, Kuwait, Lebanon, Libya, Malta, Morocco, Oman, Qatar, Saudi Arabia, Syria, Tunisia, the United Arab Emirates, the West Bank and Gaza and Yemen.
2. Excluding Iraq GDP.
3. The GCC consists of Bahrain, Kuwait, Oman, Qatar, Saudi Arabia and the United Arab Emirates.
4. According to Energy Information Administration estimates: http://www.eia.doe.gov
5. The Brent crude price was about $65/barrel.
6. See Rapport d'information du Sénat, 2007.
7. See Aoun (2008) for details about this calculation.
8. GDP per capita, PPP (constant 2000 international dollars), according to the World Bank's World Development Indicators
9. Excluding 1991–5 for Kuwait.
10. See Chapter 4.
11. Cf. Human Development Report 2007/2008: http://hdr.undp.org/en/
12. According to data from IMF (2007) and International Labour Organization (ILO) http://laborsta.ilo.org/
13. See Kaufman et al. (2008).
14. World Bank, World Development Indicators 2007.
15. Public sector employment is estimated to account for almost 27 per cent of total employment worldwide, 18 per cent excluding China (World Bank 2005).
16. For more information about the disappointing experience of the oil fund in Chad, see Horta et al. (2007).
17. In contrast to the Norwegian Petroleum Fund whose investment strategy is completely transparent: http://www.norges-bank.no/Templates/Article 69365.aspx
18. The United Arab Emirates is a federation of seven states situated in the southeast of the Arabian Peninsula in South West Asia on the Persian Gulf. The seven emirates are Abu Dhabi, Ajman, Dubai, Fujairah, Ras al-Khaimah, Sharjah and Umm al-Quwain. The population of the UAE is 4.5 million inhabitants (of which about 1.5 million are Indians). The political structure of the

Federation gives independence to individual Emirates in the management of their resources. The rulers of each Emirate (emirs) form the Supreme Council which ratifies the laws, defines the orientation of the general policy of the federation and elects a president and a vice-president for a period of five years.

19. See Fonda (2006).
20. According to Human Rights Watch, in 2005, there were 2,738,000 migrant workers in the UAE, who make up 95 per cent of the UAE workforce in the private sector.

References

Al-Moneef, M. (2006) 'The Contribution of the Oil Sector to Arab Economic Development', *The OPEC Fund for International Development (OFID)*, Pamphlet Series 34, Vienna, Austria, September.

Alnasrawi, A. (2002), *Iraq's Burdens: Oil, Sanctions and Underdevelopment*. London: Greenwood Publishing Group.

Aoun, M. C. (2008) 'La rente pétrolière et le développement économique des pays exportateurs'. PhD dissertation, Université Paris-Dauphine.

Askari, H. and Jaber, M. (1999) 'Oil-exporting Countries of the Persian Gulf: What Happened to All That Money?' *Journal of Energy Finance and Development*, 4: 185–218.

Auty, R. M. (2001) 'A Growth Collapse with High Rent Point Resources: Saudi Arabia', in R. Auty (ed.), *Resource Abundance and Economic Development*. Oxford and New York: Oxford University Press, pp. 193–207.

Belayachi, K. O. and Haidar, J. I. (2008) 'Competitiveness from Innovation, Not Inheritance', in *Celebrating Reform 2008: Doing Business Case Studies*. Washington, DC: World Bank.

Birdsall, N. and Subramanian, A. (2004) 'Saving Iraq from its Oil', *Foreign Affairs*, 83, 4: 77–89.

Bolt, K., Matete, M. and Clemens M. (2002) 'Manual for Calculating Adjusted Net Savings', World Bank, Environment Department, September.

Chevalier, J. M. (1973) *The New Oil Stakes*. New York: Beekman Books Inc.

Chevalier J. M. and Aoun, M. C. (2007) 'Geopolitics of Oil and Gas Exporting Countries', in *Encyclopaedia of Hydrocarbons*, ENI-TRECCANI, Vol. IV.

Collier, P. and Hoeffler, A. (2004) 'Greed and Grievance in Civil War', *Oxford Economic Papers*, 56: 563–95.

Corden, W. M. and Neary, J. P. (1982) 'Booming Sector and Deindustrialization in a Small Economy', *Economic Journal*, 92, 368: 825–48.

Cordesman, A. H. (2006) 'Saudi Energy Security: a Global Security Perspective', paper presented at the Chatham House conference 'Investment in Middle East Oil: What is at Stake?' November.

Davis, J., Ossowski, R., Daniel, J. A. and Barnett, S. (2003) 'Stabilization and Savings Funds for Non-renewable Resources: Experience and Fiscal Policy Implications', in J. M. Davis, R. Ossowski and A. Fedelino (eds), *Fiscal Policy Formulation and Implementation in Oil Producing Countries*. International Monetary Fund, pp. 273–315.

Eifert, B., Gelb, A. and Tallroth, N. B. (2002) 'The Political Economy of Fiscal and Economic Management in Oil-exporting Countries', *World Bank Policy Research Working Paper 2899*, October.

Fasano, U. (2000) 'Review of the Experience with Oil Stabilization and Savings Funds in Selected Countries', *IMF Working Paper*, WP/00/112, June.

Fasano, U. and Iqbal, Z. (2003) *GCC Countries: From Oil Dependence to Diversification*. International Monetary Fund.

Fonda, D. (2006) 'Inside Dubai Inc.', *The Times*, 6 March.

Horta, K., Nguiffo, S. and Djiraibe, D. (2007) 'The Chad-Cameroon Oil & Pipeline Project: a Project Non-Completion Report', Environmental Defense, ATPDH, CED, April.

Hotelling, H. (1931) 'The Economics of Exhaustible Resources', *Journal of Political Economy*, 39, 2: 137–75.

International Energy Agency (2005) *World Energy Outlook: Middle East and North Africa Insights*.

International Monetary Fund (2005) 'United Arab Emirates: Selected Issues and Statistical Appendix', *IMF Country Report*, No. 05/268, August.

International Monetary Fund (2007) *Guide on Resource Revenue Transparency*, June.

International Monetary Fund (2008) 'Regional Economic Outlook: Middle East and Central Asia', *World Economic and Financial Survey*, May.

Karl, T. L. (1999) 'The Perils of the Petro-state: Reflections on the Paradox of Plenty', *Journal of International Affairs*, 53, 1: 31–48.

Kaufman, D., Kraay, A. and Mastruzzi, M. (2008) 'Governance Matters VII: Aggregate and Individual Governance Indicators 1996–2007', *Policy Research Working Paper 4654*, Washington, DC: World Bank.

Kurtzman, J., Yago, G. and Phumiwasana, T. (2004) 'The Global Costs of Opacity: Measuring Business and Investment Risk Worldwide', *MIT Sloan Management Review*, October.

Leite, C. and Weidmann, J. (1999) 'Does Mother Nature Corrupt? Natural Resources, Corruption and Economic Growth', *IMF Working Paper* WP/99/85.

O'Higgins, E. R. (2006) 'Corruption, Underdevelopment and Extractive Resource Industries: Addressing the Vicious Cycle', *Business Ethics Quarterly*, 16, 2: 235–54.

Otto, J., Andrews, C., Cawood, F., Dogget, M., Guj, P., Stermole, F. and Tilton, J. (2006) *Mining Royalties: a Global Study of their Impact on Investors, Government and Civil Society*. Washington, DC: International Bank for Reconstruction and Development, World Bank.

Ricardo, D. (1821) *On the Principles of Political Economy and Taxation*. Third edition. London: John Murray.

Ross, M. (2001) 'Does Oil Hinder Democracy?' *World Politics*, 53: 325–61.

Sénat (2007) 'Le nouvel "âge d'or" des fonds souverains au Moyen-Orient', Rapport d'information no. 33, by J. Arthuis, P. Marini, A. de Montesquiou, P. Adnont, M. Moreigne and P. Dallier, 17 October.

Setser, B. (2007) 'Oil and Global Adjustment', Peter G. Peterson Institute for International Economics, March.

Setser, B. and Ziemba, R. (2007) 'Understanding the New Financial Superpower: the Management of GCC Official Foreign Assets', *RGE Monitor*, December.

Torvik, R. (2001) 'Learning by Doing and the Dutch Disease', *European Economic Review*, 45: 285–306.

Truman, E. M. (2007) 'Sovereign Wealth Funds: the Need for Greater Transparency and Accountability', *Policy Brief* No. PB07-6, Peter G. Peterson Institute for International Economics, August.

World Bank (2005) *Middle East and North Africa Economic Developments and Prospects 2005: Oil Booms and Revenue Management.*

Yashir, F. (1988) *Mining in Africa Today: Strategies and Prospects*. London: Zed Books and The United Nations University.

Zhang, Y. (2008) 'Opportunities for Mitigating the Environmental Impact of Energy Use in the Middle East and North Africa Region', *Discussion Paper*, ESMAP, March.

6
The United States Energy Policy: At a Turning Point

Sophie Méritet and Fabienne Salaün

1 Introduction

The exceptional development of American capitalism was founded upon abundant, cheap and domestic energy resources: coal, oil, natural gas, hydroelectricity and nuclear. The abundance, low prices and low taxes did not encourage energy efficiency. The United States consumes roughly 70 per cent more energy per capita or per dollar of GDP than most other developed countries. The country, which represents 5 per cent of the world's population accounts for 25 per cent of the world's energy consumption.

After a long period of energy self-sufficiency, the United States is now importing more and more crude oil, petroleum products and natural gas. Growing energy dependence raises the question of security of supply in a world where oil and gas resources are concentrated in a few countries subject to geopolitical turbulences. The country is competing with Asia and Europe to get access to resources.

Another dimension of the energy question in the United States is the organisation of the natural gas and electric power industries which have been a subject of debate and controversy since the late 1970s (Kahn, 1988). Joskow and Schmalensee (1983) warned us that the process will take time: 'If deregulation is to play a role in helping to improve the efficiency with which electricity is produced and used, it must be introduced as part of a long-term process that also encompasses regulatory and structural reform.'

Lastly, there is the question of climate change which has long been ignored by both the administration and industry. The United States has not ratified the Kyoto Protocol but we will see that the picture is now changing. More and more people, states and local communities are becoming concerned by this issue.

This chapter is divided into two main parts underlying the energy and climate change challenges faced by this economic superpower. Section 2 presents the energy situation with the increasing and worrying dependence on energy imports, and discusses the lessons to be learned from the liberalisation of the electricity and natural gas markets. Section 3 is focused on the upcoming challenge of climate change and how the United States is going to address it.

2 The American energy situation: past, present and future

2.1 An increasing and worrying dependence on energy imports

Historically, energy has been abundant and cheap in the United States. American economic growth and welfare relied on it. The US has been blessed with large endowments of domestic energy resources: oil, natural gas, coal and hydropower. The country was self-sufficient but the situation has evolved. Domestic resources are depleting or protected by environmental constraints. The energy dependence from imports grows rapidly for oil and natural gas. The question of security of energy supply is highly strategic.

2.1.1 The current energy balance

In 2006, about 39 per cent of US energy consumption came from oil, followed by 23 per cent from natural gas, 23 per cent from coal, 8 per cent from nuclear power and 7 per cent from renewable energy (primarily conventional hydroelectricity resources). The mix has changed little since 1973 (Figure 6.1)

Coal. The United States has the world's largest proven coal reserves (27.1 per cent of the world reserves at the end of 2006). The country is the seventh exporter in the world with more than 40 Mt in 2006. Historically, coal has been a source of cheap energy. The abundance of this resource and improvement of productivity for mining and transportation have ensured that the price has remained low compared to other energy resources. Low and stable prices make coal ideal for power generation: coal accounts for 50 per cent in power generation which absorbs around 90 per cent of the American coal production. Despite the implementation of the Clean Air Act of 1970, coal generation has nearly tripled since that date.[1] In the ten states where coal is the most used, electricity rates are 40 per cent lower than in the ten states that use other fuels.

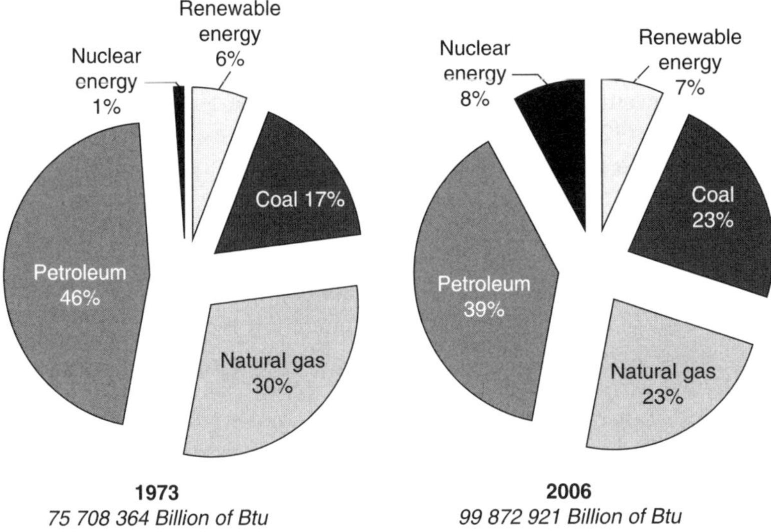

Figure 6.1 Primary energy consumption by sources in 1973 and in 2006 (in %)
Source: Energy Information Administration.

Coal is the most polluting fossil fuel even if polluting emissions have been cut by one-third since 1970. Considering the coal share in the American energy mix, funds were voted in the Bush energy programme to develop new clean coal technologies. Looking at the resistance of some non-governmental organisations to any systems that do not have complete carbon capture and sequestration, there will be a great deal of work ahead for industry and government. Coal's relative abundance and the security the vast domestic reserves provide suggest that coal could continue to play a key role in meeting American energy needs well into the future if coal is used in as efficient and as clean a manner as possible. At the same time, energy companies are yielding to environmental demands: for example, the private equity fund purchasing Texas Utilities (TXU) obtained the dropping of TXU's applications for eight proposed coal plants in Texas, and also many other commitments to reduce air pollution and global warming emissions. The carbon risk associated with mergers now warrants significant concern from investors.

Natural gas. Ever since the Second World War the United States has always relied more on this fossil energy than other developed countries. Natural gas constituted around a quarter of the nation's total primary energy supply in 2007. According to the Department of Energy (DoE (2008)), consumption growth in the United States will be led by utilities that require natural gas to produce electricity (28.1 per cent) and industrial (30 per cent), followed by commercial (13.1 per cent) and household consumption (20 per cent). Total consumption is expected to grow by 9 per cent through 2025. Natural gas holds an important place in the American electric market as the second largest source of fuel, after coal, and the fastest-growing fuel for power generation. Around 19 per cent of electricity comes from gas fired plants, up from only 10 per cent in 1986 (when gas prices were fully deregulated). Natural gas accounts for 90 per cent of all new electric power generation capacities installed in the United States since 1995. It has also become a popular fuel among residential users for heating and cooking. More than half of all households today are heated by natural gas. While some studies have predicted a decrease in gas demand resulting from the development of clean coal technologies and a resurgence in nuclear power, there is still a general agreement that nuclear power will not be a near-term solution, and there is also a disagreement over how soon carbon capture might be available as a commercially proven technology.

Until the late 1990s, the North American market for natural gas was essentially independent from other major gas markets, even if the United States has been importing gas since the end of the 1950s. In this regional market, the US is by far the largest consumer of natural gas, accounting for 85 per cent, followed by Canada at 10 per cent, and Mexico at 5 per cent. American production has not increased sufficiently to meet the rising demand. Over the last few decades, the gap has been met by boosting imports. The volume of net natural gas imports equalled 16 per cent of gas consumption, a ratio that has remained relatively stable in the past ten years. In 2006, Canada remained the largest natural gas exporter to the United States, but the imports in volumes are decreasing. Most of the imports (85.7 per cent in 2006) arrived by pipelines from Canada. Natural gas imports are expected to grow in the future as demand for gas continues to rise and production in this region is not expected to keep pace. Both the Canadian and United States natural gas fields are maturing. Moreover, twenty years ago, nearly 75 per cent of federal lands were available for private energy exploration companies. Since then, the share has fallen to 17 per cent. Environmental and land-use considerations have prompted the administration to remove land from energy development that was available for exploration. Today, the question of drilling in ecologically protected areas has generated much debate particularly around Alaskan preserves and Arctic National Wildlife

Refuge Development Issues, the use of which the United States Congress approved.

Depending on the level of natural gas prices, domestic production has been forecasted to increase roughly 50 per cent by 2020 after increasing only 8 per cent during the 1990s. Imports primarily from Canada are estimated to increase by about 80 per cent and would account for nearly 20 per cent of the United States natural gas supply. These forecasts reflect the assumption that higher prices will lead to significant increases in resources, production from offshore and onshore, and non-conventional non-associated sources. Additional gas from the less accessible areas of Canada as well as the long planned gas pipeline from Alaska to 'the lower 48 states' could increase supply.

Even if North American production continues to play a major role, liquefied natural gas (LNG) imports from overseas are expected to be the most probable supply source to meet future increases in consumption. In 1986 LNG imports represented 0 per cent but in 2006 they represented 3 per cent of the total natural gas consumption. Moreover, with the construction of new facilities, imports are expected to rise substantially in the coming years. Therefore, it will diversify the American gas supplies. Many analysts predict that only twelve of the forty LNG terminals being considered will ever be built because of environmental constraints, associations' actions and other factors.

The United States' natural gas imports are expected to rise significantly in the next two decades, raising concerns about supply security. Over time, there will be growing pressure on the US to develop the capacity to manage disruptions to gas supplies. The best protection against those insecurities is to sustain the North American gas production base. It is also necessary to integrate the American natural gas market into the other markets through a robust network of LNG terminals, pipeline transportation and gas storage.

Oil. The oil situation is more worrying. Americans are the largest consumers of oil in the world. The United States consumes about 20.6 Mb/day or roughly 25 per cent of global demand[2] which equals the combined consumption of the five largest national consumers (China, Japan, Germany, Russia and India). Currently per capita, American oil consumption amounts to 68.4 barrels of oil per thousand people per day compared to the Chinese demand which only stands at 5.5 barrels per thousand people per day. The United States is the third largest producer of oil in the world behind Saudi Arabia and Russia, having produced 8.3 Mb/d of crude oil in 2006. Expanding production in the United States is difficult due to a combination of geological parameters, economics and policy issues. Existing conventional oil that is produced in America comes from mature regions that have been fairly well explored and are now experiencing geologically driven production declines. Recently, technology and higher prices have made it profitable to explore in deeper waters in the Gulf of Mexico, and new oil reserves are being discovered but the process of exploration and development can take years before oil is actually produced.

In the State of the Union address in 2008, President Bush admitted that America had a problem: *oil addiction*. The demand for oil has increased so steadily over the last few decades that it has led to a greater dependence on imported oil. The United States is more dependent on foreign oil than ever before. In 1948, it became a net importer of oil. Imports of oil increased rapidly in the early 1970s: 35 per cent of total demand in 1973 and 60 per cent in 2006. It is important to underline that in 2008, the demand for oil in the United States declined slightly for the first time. This is what Yergin (2008) calls the 'peak demand'. Nevertheless, the United States remains the top importer of oil in the world. Many analysts forecast that petroleum imports will continue to grow. By 2020 these imports will represent 75 per cent of the total US petroleum consumption. More recently, America has even begun to import more and more refined oil products as well. This is in part due to demand but more so because domestic oil refining capabilities have not kept pace with rising consumption. Operable refinery capacity has steadily increased at an average annual rate of 1.1 per cent over the last ten years from approximately 15.6 Mb/d in 1997 to 17.4 Mb/d in 2007. However, in 2008, refinery utilisation rates (81.4 per cent) have been abnormally low. This production slowdown in refineries, now operating at levels last seen in the aftermath of Hurricane Katrina, could not have come at a more troublesome time with the high price of oil and the falling value of the US dollar. In 2005, Hurricanes Rita and Katrina caused refineries to cut more than 23 per cent of their total oil refining capacity to 17.1 Mb/d. Since these two hurricanes, the administration has been focusing on the depleting stocks of oil products. At the same time, the gap between the price of crude oil and the price of gasoline and other refined products has put pressure on refiners' profit margins.

American dependence on imported crude oil and products derives primarily from the transportation sector. The US demand for oil revolves around four major sectors: transportation (70 per cent), industry (24 per cent), electricity generation (2 per cent) and residential/commercial needs (5 per cent). Transportation represents the main share of oil consumption of which two-thirds is motor gasoline. American consumers are accustomed to cheap and plentiful gasoline and have structured cities and lifestyles around this fact. They own more than 242 million motor vehicles, close to one vehicle per person. US road petroleum use represents 33 per cent of all such use globally which is twice as high in percentage terms as the whole of Europe. Given that oil is used in the transportation sector it will be very difficult for alternative energy to replace imported oil in the next few years. Current ethanol production is not yet significantly replacing gasoline. It is targeted to reach 15.2 billion gallons a year or close to 1 Mb/d by 2012 under the new 2007 Energy Independence and Security Act. However, continuing to grow domestic ethanol production at this pace over the next five to ten years will prove highly challenging as food and other agricultural prices have skyrocketed this past year in response to the new demand for corn. Moreover, it will only eliminate future increases in oil imports, not actually lower them from the current level.

Electricity. The electricity mix of the United States is currently dominated by coal-fired plants which generate approximately 50 per cent of the power and represent 31 per cent of the total installed capacity (see Figure 6.2). Nuclear power plants are the second major source of generation (20 per cent of the generation and 10 per cent of the installed capacity); this technology was first developed in the United States where 104 reactors are operated. The first oil shock boosted the development of this technology, before a significant slowdown at the end of the 1970s, due to economic factors (overestimation in demand forecasting, inflation of costs due to the delays in construction) and also to the accident at 'Three Mile Island' in 1979. The latest order for a nuclear power plant dates from 1977 (Seebrook in New Hampshire which has been operational since 1990). More recently, the significant improvements in the performance and competitiveness of these plants have provided new advantages to nuclear energy. The 'renaissance' of nuclear energy in the United States is manifest today and the Energy Policy Act of 2005 was clearly oriented to support it.

Since 1973, the use of electricity has increased by 66 per cent, somewhat less than the GDP at 80 per cent. While most of the electricity is still being generated from coal, major challenges will have to be faced in this sector which is also one of the principal sources of CO_2 emissions. Even with the significant shift from oil to nuclear since 1973 (respectively

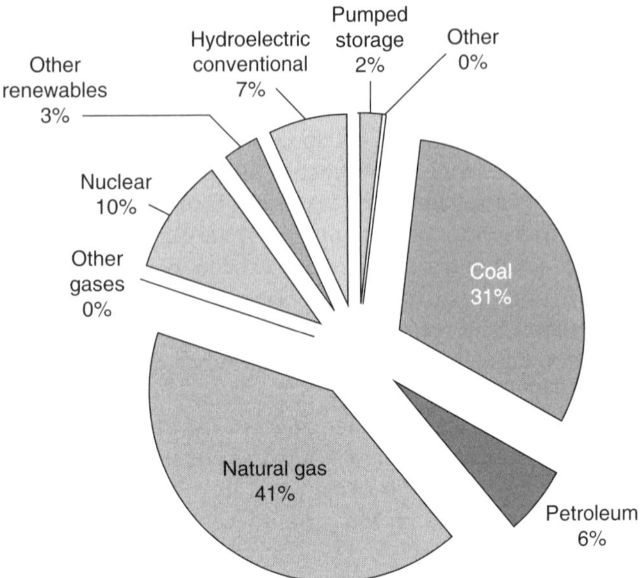

Figure 6.2 Total installed capacity by energy source in 2006 (in %)
Source: Energy Information Administration.

about 17 per cent and 5 per cent of electricity at this period against 3 per cent and 20 per cent in 2006), the electricity generation mix in the United States will still be dominated by coal in 2030 (see Figure 6.3). The strategic issue of security of supply could favour the use of the domestic coal resources.

The United States will need 44 per cent more electricity by 2020 to meet its growing energy demand. While natural gas fired power plants will provide most of the capacity additions needed by 2015/16, more coal fired plants will be built in the next few years. The natural gas share will fall to 14 per cent in 2030 and the coal share will increase to 54 per cent. With a 32 per cent increase between 2006 and 2030, the share of renewable energy sources will jump to 13 per cent of the total electricity supply. Federal tax incentives, state renewable energy source programmes and rising fossil fuel prices will support this significant increase. At the same time, nuclear generation will increase thanks to the capacity additions and to improvements in the performance of existing nuclear facilities. Nevertheless the share of nuclear will not significantly change (from 20 to 18 per cent).

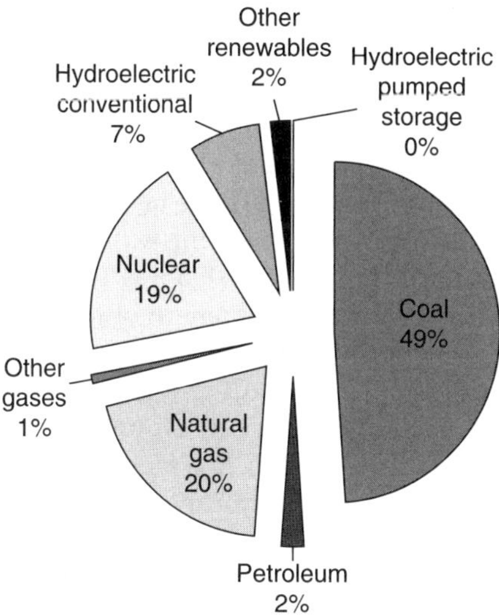

Figure 6.3 Total generation by energy source in 2006 (in %)
Source: Energy Information Administration.

Several reasons explain these evolutions. The growing share of natural gas has been supported by the deregulation of energy industries: due to the high capital intensity of nuclear and coal plants, gas has been preferred by investors, especially independent power producers. However, the rising prices of fossil energies have begun to deter investment in gas fired plants. Federal incentives for new nuclear capacities will support investments in new facilities and the abundance of domestic coal will help towards future energy independence which is a major issue for the United States.

The major challenge for the United States in the next few years will be to solve the following equation: how to satisfy the growing demand for electricity while dealing with climate change. The electricity industry will have a major role to play either through the technologies portfolio that will emerge in supplying consumers or in the field of demand side management.

2.1.2 What does security of supply mean today for the United States?

Affordable and reliable supplies of energy play an important role in fostering economic growth and development. American economic growth

Box 6.1 Nuclear energy in the United States

The United States has the world's largest nuclear power generating capacity (100 GWe). They are all light water reactors, a technology invented and initially developed in the United States in the 1950s. Their operation during recent years has been excellent. In 2007, an average capacity factor of 91.8 per cent was achieved. The average operating cost of nuclear plants in 2006 was 1.66 cents per kilowatt-hour. However, the dynamic development of nuclear capacity in the United States during the 1960s and 1970s was stopped after the Three Mile Island accident in 1979, and there have been no new reactor orders since then. The reasons for this evolution were large cost and schedule overruns in construction (for most projects, by a factor of two or three compared to original estimates). This poor performance was linked to: a change in safety regulation, forcing design changes during construction; lawsuits by opponents; and complete lack of standardisation. There were many owners, many vendors, and many architect-engineers, so that no more than four of the total 104 units were identical. However, the technology itself was sound; licensed to and further developed in other countries (France, Japan, South Korea) it has been and continues to be a technical and economical success story. US nuclear utilities are continuing to consolidate: in a few years, only ten or twelve utilities are expected to be involved in nuclear generation, versus 101 in 1991. Operating licences have been extended from 40 to 60 years for 48 reactors up to now, and some 37 more licence extensions are anticipated.

The government and Congress, concerned by security of energy supply and CO_2 emissions, decided to help revive nuclear power in the United States and enacted the Energy Policy Act of 2005. Incentives in this act include, notably, a production credit of 1.8 cents per kilowatt-hour for the first six GWe of new nuclear capacity over the first eight years of operation; federal risk insurance totalling $2 billion to compensate for regulatory delays; and federal loan guarantees covering up to 80 per cent of the total project cost. In addition, the Nuclear Regulatory Commission developed a new, more predictable licensing process that includes generic design certification (favourable to standardisation), early site permits (independent of reactor design), and a combined construction–operating licence, COL. These processes are long – typically two to four years each – but they can be done independently and at least partly in parallel. However, legal challenges could still lead to delays. Thanks to these measures, by autumn 2008 there are COL applications covering fifteen reactors totalling 18 GWe, plus 23.5 GWe expected before mid-2009. A COL application does not oblige an applicant to build a reactor. On the other hand, preparing a COL application is expensive and time-consuming and therefore indicates some degree of commitment.

Five types of reactors are proposed in the US by the main international vendors. How many reactors will be built? The most optimistic projections see an initial COL award in 2010. Probably at least 6 GWe of nuclear capacity will be under construction by 2010–12 to take full benefit of the government subsidy. According to recent utility cost projections, without government subsidies for nuclear or significant penalties for CO_2 emissions, coal is cheaper than nuclear. On the other hand, public opinion is more favourable to nuclear than to coal: a May 2008 poll indicated that if a power plant were built in their community, 43 per cent of residents preferred nuclear, 26 per cent gas, and only 8 per cent coal.

For the longer term, dynamic development of nuclear energy will oblige the United States to resolve the problems of nuclear waste management and uranium supply. The Yucca Mountain repository planned in Nevada is strongly opposed by state politicians and is still far from completion. Moreover, as planned this repository could serve only existing reactors. The government is considering introducing reprocessing of spent fuel and 'burning' waste products in a fast neutron reactor, a solution which would decrease by a large factor the toxicity of final waste. In case of uranium market tensions, fast reactors could be operated as breeders.

C. Pierre Zaleski (CGEMP).

and welfare have relied for decades on abundant and relatively cheap energy. The 'American way of life' also relied on it. Distortions in prices, consumption, supply or reliability of energy infrastructure services can lead to large economic and social costs. A growing fraction of the United States energy consumption is supplied by imports of energy, primarily petroleum from other countries. More and more natural gas is also imported. Recent power blackouts have raised the question of the reliability of the electricity system. This poses a new kind of threat to national security and prosperity in the United States as President Bush emphasised.

The security of supply issues can be raised in different ways. Some oil and gas reserves are concentrated in areas which are politically unstable and have governments that are not always friendly to the United States. The authorities link their energy policy to their foreign policy, especially when oil is in the middle of the discussions. American diplomatic relations affect decisions that influence the price of oil. Another answer is to develop the national production of fossil fuels to foster the domestic supply: coal, oil and gas. For example, high prices made it economically competitive to exploit some reserves in difficult geographical areas. One of the United States authority's responses to the security of supply issue and the growing energy demand is to develop its national energy supply. The United States is looking for additional supplies in two ways. First, protected areas all over the country are now being opened to prospecting and in some areas, prospecting is accelerating (offshore, Alaska, California, etc.). Second, outside the country, the United States is looking to obtain oil through its foreign policy as in Mexico. The invasion of Iraq can be perceived as a means to gain access to oil. It does not seem realistic to believe that the domestic supply-side initiatives will halt the increase in demand. Some demand-side policies could be another answer to security of supply. For instance, the United States is increasing its use of renewable fuels and reducing its dependence on oil through improved energy efficiency. In December 2007, President Bush signed the Energy Independence and Security Act to improve vehicle fuel economy, to increase the use of alternative fuel and to financially support alternative energy sources.

In his State of the Union address in January 2008, President Bush announced an ambitious plan to strengthen America's energy security. He laid out a plan that builds energy security by promoting diversity of both the types and sources of energy. The first two parts of the plan are the diversity of energy sources and the wise management of energy demand. A way forward for both is through new technology. A centrepiece of the plan is to reduce America's gasoline usage by 20 per cent in ten years. To achieve this, the United States needs to diversify the fuels used to power cars and trucks by increasing the use of renewable biofuels, like ethanol, and by using energy more wisely by setting high standards for automotive efficiency. The third and fourth parts of the plan for energy security will reduce, over time, the nation's oil import dependency. The plan calls for doubling the size of strategic petroleum reserves and stepping-up the production of domestic oil supply in environmentally sensitive ways. It is clearly a policy of supply. Demand-side policies could lessen the energy demand without changing the American pattern of consumption. The question is: is it sustainable?

2.2 Market liberalisation for the American energy industries: what lessons can be learned?

The transition to competitive wholesale and retail markets for electricity and natural gas in the United States has been a difficult and contentious process. Significant progress has been made especially on the wholesale competition front but major challenges must still be tackled. The North American natural gas and electric power markets offer us (Europeans) an amazing experience that can be used to build a common energy market.

2.2.1 Federal versus state level intervention

The regulation of the energy sectors in the United States is managed at the federal and at the state levels. The Federal Energy Regulatory Commission (FERC), an independent agency, regulates essentially interstate issues for the transmission of electricity, natural gas and oil. In addition, FERC reviews proposals to build liquefied natural gas terminals as well as licensing hydropower projects. The role of infrastructures such as high voltage transmission lines, oil and gas pipelines and liquefied natural gas terminals are now generally classified as 'essential facilities' that play a key role in ensuring security of supply. Hence, the Energy Policy Act of 2005 has empowered FERC:

- To promote the development of a strong energy infrastructure. FERC has full authority to backstop states in siting transmission lines used for interstate commerce in corridors designated by the Department of Energy as being in the 'national interest'. However, states retain primary jurisdiction and FERC only becomes involved in certain situations where states do not or cannot site the facilities.
- To support competitive markets. The Orders 888 and 889[3] have been completed to authorise FERC to reform the system of transmission tariffs and to continue to support the creation of Regional Transmission Organisations (RTOs) and independent system operators (ISOs) which constitute an essential part of the liberalisation process. The FERC also has prerogatives in the field of merger and acquisition review, an issue which is becoming crucial in the new context of ISOs and the RTOs in which energy companies can develop either an organic or external growth of their activities.

Box 6.2 Energy and environmental laws in the United States

1920: Federal Power Act (FPA)
1935: Public Utility Holding Company Act of 1935 (PUHCA)
1938: Natural Gas Act
1959: National Environmental Policy Act (NEPA)
1963: Clean Air Act (amendments in 1977 and 1990)
1972: Coastal Zone Management Act
1977: Clean Air Act Amendments and Clean Water Act
1978: Natural Gas Policy Act and Public Utility Regulatory Policies Act (PURPA)
1989: Natural Gas Wellhead Decontrol Act
1992: Energy Policy Act
1997: Orders 888 and 889
2005: Energy Policy Act
2007: Energy Independence and Security Act

FERC focuses on and is limited to the regulation of interstate issues while other bodies intervene at the level of the states. In each state, a public utilities commission (PUC) is in charge of regulating the rates and services of all public utilities, i.e. of all industries which are characterised by the existence of a natural monopoly. The coexistence of these two levels of regulation is not always trouble-free and problems can arise between FERC and some states. The question raised by this two-level regulatory framework also exists in the case of environmental issues as will be seen below.

Natural gas. The regulation of the natural gas industry in the United States has historically been a bumpy ride, leading to far-reaching changes in the industry over the past thirty years. In the past, gas producers explored for and produced natural gas and sold it to pipeline companies. The pipeline companies transported the gas across the country and sold it to local gas utilities. The local utilities sold and distributed the gas to their customers. The federal government regulated the price at which producers could sell their gas to interstate pipeline companies. It also regulated the price at which pipeline companies could sell the gas to local utilities. Interstate pipelines acted as both transporters as well as sellers of the commodity, both of which were rolled up into a bundled product and sold for one price. State agencies regulated the price that local gas utilities could charge customers. In the last few decades the industry has undergone major changes, spurred by the ever-changing regulatory environment. The current regulatory environment is much less stringent and relies more heavily upon competitive forces. Despite the deregulation of some activities, regulatory forces still keep a watchful eye over the industry especially in the transportation and distribution sectors.

Today, gas producers and marketers are not directly regulated. It does not mean that there are no rules; it just means that there is no government agency charged with the direct overseeing of their daily business. However, the prices they charge are linked to competitive markets and are no longer regulated. In the United States the average producer of gas is a small company. The production side is not the only one subjected to intense competition; the buyers' side of the wellhead market is too. This can be explained by the fact that many gas fields are reachable by more than one pipeline and as pipelines are extensively interconnected, it is clear that the market is extremely competitive.

Under the current regulation, only pipelines and local distribution companies (LDCs) are directly regulated with respect to the services they provide. Interstate pipelines, usually owned by entities other than producers, link wells with consuming areas. They are still regulated in the rates they charge, the access they offer to their pipelines, and in the siting and construction of new pipelines. They can serve only as transporters of natural gas and are under the authority of the FERC. Pipelines must now offer access to their transportation infrastructure to all other market players equally. Despite criticism, these were the primary forces that moved the gas industry to competition. By converting pipelines into transporters customers are able to search for attractive purchases, rather than being forced to take whatever gas the pipeline chooses to buy.

LDCs are regulated by state utility commissions which oversee their rates and construction issues. They ensure that a proper procedure exists and check that they maintain adequate supply to their customers. The regulation of distribution is currently undergoing a process of change with the adoption by many states of procedures aimed at exploring and setting up retail choice programmes. These programmes allow consumers more flexibility in arranging for the delivery of their gas, including allowing many customers the option of purchasing their own natural gas, and using the distribution network of their LDC simply to transport gas. By early 1998, local gas utilities in eighteen states and the District of Columbia had proposed or implemented residential customer choice programmes, giving 13.3 million households, or a quarter of all residential natural gas customers nationwide, the chance to select their natural gas supplier by the year 2000. While it has been proven that commercial and industrial customers have benefited from competition, there is no guarantee that residential customers will save money by purchasing natural gas from a non-utility supplier.

The deregulation of natural gas markets in the United States is often perceived as a success story. Gas prices fell with the introduction of competition. They are now more volatile, however. The natural gas market is liquid (more so than in the European Union) and most of the transactions are operated in the spot markets (and not through long-term contracts). The level of competition is high and involves a large number of actors.

Electricity. After several decades of stability, the power sector has undergone major changes since the 1970s. The Public Utility Holding Company Act (PUHCA) was enacted in 1935 to protect investors and consumers from the abuses of utility holding companies. Under PUHCA, any company that 'owns, controls, or holds with power to vote' 10 per cent or more of the voting stock of an electric utility is deemed a 'holding company'. It imposes numerous ownership limitations and reporting requirements to the Securities and Exchange Commission (SEC). More precisely, PUHCA prohibited the merger of geographically diverse utilities, and limited diversification to those potential acquisition targets that supply or serve the utility. For this reason, the electric power sector in the US has been organised for a long time either at the level of states or municipalities with a strict regulation of tariffs. At the end of the 1980s, several problems emerged due to the low level of regulated tariffs in some states that did not cover the cost. It was the period of the 'negawatts' programmes[4] launched by several companies keen to foster energy efficiency rather than investments hence avoiding additional financial losses. The efficiency of the regulation of the power sector has been called into question, especially in California and the north-east where the price of electricity is higher than in other states. It has led big consumers to threaten the states to leave for other parts of the country. In 1992, the Energy Policy Act (EPA) opened the door for the first time to independent power producers and traders to negotiate on market-based rates. In 1996, the Orders 888 and 889 took this a step further by organising the opening-up of wholesale power markets in order to favour competition and exchanges between states. At the same time, the opening-up of retail markets has continued to be a prerogative of states and most of them have stopped or delayed the deregulation after the Californian crisis.

The Californian experiment was the first one on such a large scale. The failure of the market design led to chaos with shortages over several days while they were trying to cope with the lack of power capacities. However, this crisis proved to be a lesson in what errors to avoid: the implementation of a compulsory pool in which all companies have to buy their electricity at market-based rates on the one hand, and to sell this power at regulated tariffs on the other. The facts revealed that such a market design was not sustainable. The lack of investment led to a tightening up on the market, without any possibility for utilities to pass on the rise of the wholesale prices to end users. The bankruptcy of these companies was unavoidable. This market design crisis[5] was followed by the Enron case but one needs to distinguish the two events: the Californian crisis can be seen as a failure of the market design, the Enron case can be seen as a business model which came too early in an immature market. Enron tried to take advantage of the opportunities offered by the deregulation of the energy markets. The business idea was to make arbitrages between the different markets and to trade commodities without owning any physical assets. Nevertheless, this business idea was carried out with little regard to ethical issues and Enron's activities were not controlled by any regulatory body. In addition, the financial regulation on special purpose entities authorised the company to organise an opaque galaxy of subsidiaries in order to hide its losses and debts. For several years, Enron was able to officially show artificial profits. In the beginning of 2001, this situation had become unsustainable and Enron's bankruptcy was unavoidable. Mismanagement as well as blatant dishonesty caused inestimable hurt and loss to the employees. The Enron case provoked a worldwide shockwave and has raised significant debates around the financial regulation and governance of firms in the United States as well as in Europe.

The Enron case as well as the Californian crisis have made most states delay or stop their deregulation programmes: the deregulation of retail markets is currently active in less than twenty states. The regulation of the wholesale market is the FERC's prerogative but states are free to choose to deregulate or not their retail markets. It is one of the specificities of the American situation where two levels of regulation are active. The opposition between these two levels can be severe: this has been the case with the attempt of the FERC to impose its 'standard market design' (SMD). This SMD was inspired by the Pennsylvania–New Jersey–Maryland (PJM) situation: PJM is the independent system operator in the interconnected network of different states. It is a model because it has implemented innovative rules in order to improve the efficiency of the

Box 6.3 The Enron case

The bankruptcy of Enron in 2001 is a symbol of the history of the liberalisation of the energy markets in the 1990s. Enron was a gas transmission company which has diversified its activities in the trading of commodities. With the opportunities offered by the liberalisation in gas and electricity markets, the business model of the company was to make profit from the spreads existing between the different markets at different times and in different locations. In addition to this new business, Enron has developed innovative financial products such as 'weather derivatives' in order to offer new tools for hedging new risks in these new energy markets. The crisis in California in 2000, another symbol of the energy markets' history, gave rise to the first doubts about the company which had been accused of manipulating the prices in the power market and thus making the crisis worse. The development of highly risky business which had been essentially based on purely financial transactions in addition to the bad use of some legal rules of accountancy permitted Enron to artificially exaggerate its profits and to hide its losses. Moreover the governance of the company had clearly failed both in terms of transparency and honesty. The management was able to misappropriate millions of dollars and was convicted of theft. Thousands of employees lost their jobs and their pensions and often also lost their savings because of the incentive for investing in shares of the company. The Enron case led Congress to reinforce the regulatory framework for financial transactions with the adoption of the Sarbanes Oxley law in 2002. The business mind of the company was clearly very bad but the business model could prefigure the future in terms of management in complex markets.

Source: Chevalier (2008).

market mechanisms. The blackout of 2003, which did not occur in the PJM area, has increased the interest in the PJM design which has implemented both a mechanism of capacity obligation and a market for capacity in order to stimulate investments in generation plants: all suppliers have to give proof that they get the volume of generation capacity to cover their peak demand, plus a security margin. In the case of failure to fulfil this obligation, suppliers have to pay a fine, a mechanism that is supposed to encourage investments. Nevertheless, PJM face some difficulties in achieving this goal and periodically change the rules in order to improve the system. This pragmatic approach makes the PJM initiative one of the most interesting experiments of the liberalisation of the electricity sector. In addition, it has inspired the FERC in its goal to favour the creation of the Regional Transmission Organisations (RTOs). These RTOs support the development of interstate exchanges and could be supported by the FERC which holds the authority for interstate interconnection capacities. Investment in transmission capacities has begun to increase and major projects are planned for developing the infrastructure.

Last but not least, the case of Texas is also interesting because it shows that deregulation does not mean no more regulation but sometimes more and in some cases better regulation. The deregulation of the retail market in Texas has been phased in over several years beginning in 2002. The Electric Reliability Council of Texas, created in 1979, was appointed for grid reliability and operations so as to ensure no one would gain an unfair advantage in the marketplace. An original regulated rate concept was implemented to govern the pricing behaviour of the former utilities: the 'price to beat'. One concern was that incumbents would undercut the prices of new entrants, preventing enough competition and protecting their monopoly position. The price to beat introduced a phase-in period from 2002 to 2007 during which a price floor was established to prevent incumbents from adopting a predatory strategy and thus allowing new entrants to become established. New entrants could charge a price below the price to beat but incumbents could not. In order to create competition in the market, the price to beat needs to be high enough to guarantee a positive margin to new entrants (e.g. above their cost) but it also needs to be reasonably low for the customers. As a result, since 2002, approximately 40 per cent of residential customers have switched from their former incumbent supplier and competition has developed. At the same time, other regulations have spurred investments in new generation plants, especially in renewable sources of energy and Texas has become the leading wind state in the United States, with one-third of the nation's total installed wind capacity. The Texas experiment clearly demonstrates that deregulation can provide gains to consumers as well as gains to the environment. With the rise of energy prices, residents have begun to reduce their power consumption, to insulate their houses or to install solar screens. Some Texas utilities are also installing advanced power meters that might enable real-time pricing in the near future. This would permit energy customers to save money by managing their consumption differently.

3 The upcoming challenge of climate change

The United States has not yet ratified the Kyoto Protocol but more and more American citizens, states and local communities are becoming concerned about environment protection. There is an obvious growing awareness of climate change. In the absence of a federal environmental policy, states are taking some initiatives in terms of emissions reduction policy. Research and development, and demand-side efficiency are the two ways to address climate change for the American administration.

3.1 The growing awareness of climate change

The first *Annual World Environment Review*, published on 5 June 2007, revealed that the American public is aware of the problem of climate change and is expecting some initiatives from their politicians to solve the different issues that it raises. To be more precise, 74 per cent of citizens in the United States are concerned about climate change, 80 per cent think their government should do more to tackle global warming, 84 per cent think that the United States is too dependent on fossil fuels, and 72 per cent think that the United States is too reliant on foreign oil. Even if the United States has refused until now to ratify the Kyoto Protocol, because 'the American way of life is not negotiable', some signs may indicate that the United States will participate in the effort to solve some of the environmental issues.

It is clear that the United States is the greatest source of greenhouse gas (GHG) emissions in the world: with approximately 25 tons of CO_2 equivalent emissions per inhabitant (TCO_2eq), the United States dominates the world ranking ahead of the EU-25 and Japan (around 11 TCO_2eq). Electric power generation in the United States is the main source of polluting emissions before the transportation sector. There is no mandatory obligation put in place in the field of emissions control, a situation that reflects the difficulty in having a political consensus for imposing any constraint on any economic sector in this country. Until now, the climate change policy of the United States has been dominated by two major dimensions: technological development and demand-side issues.

Nevertheless, public opinion is now in favour of a positive evolution: rising and high energy prices and environmental concerns have brought energy policy to the forefront of many debates especially in the 2008 election year in the United States. The National Commission on Energy Policy recommended in 2004 that the United States adopt a mandatory economy energy programme to limit future greenhouse gas emissions in a manner that does not harm the economy. It is convinced that a combination of market signals and technology policies (for example, R&D investments) is the most promising and ultimately most effective path forward. During the debate on the Energy Policy Act of 2005, Congress voted on but did not enact[6] legislation that would cap GHG emissions. It did not require the establishment of a US climate change strategy, the reporting and disclosure of GHG emissions, the promotion of renewable energy sources and energy efficiency, the promotion of carbon sequestration, 'clean coal' power plants or automotive fuel efficiency. Nor did it support US participation in international climate change negotiations. It would have been a tremendous federal step.

The main answer of the United States to climate change is research and development (Table 6.1). Over the past decade, the United States has invested $18 billion, spending more than the EU-15 and Japan

Table 6.1 R&D budget, 2007 and 2009 (in millions of dollars)

	FY 2007 Actual	*FY 2008 Estimate*	***FY 2009 Budget***	*Change FY 08–09*	
				Amount	*Percentage*
R&D in the fiscal year 2009 Budget by agency (budget authority in millions of dollars)					
Total R&D (Conduct and Facilities)					
Defence (military)	79 009	77 782	**80 688**	2 906	3.70%
Health and Human Services	29 621	29 816	**29 973**	157	0.50%
NASA	11 582	12 188	**12 780**	592	4.90%
Energy	9 035	9 661	**10 519**	858	8.90%
Atomic Energy Defence R&D	*3 649*	*3 718*	***3 825***	*107*	*2.90%*
Office of Science	*3 560*	*3 574*	***4 314***	*740*	*20.70%*
Energy R&D	*1 826*	*2 369*	***2 380***	*11*	*0.50%*
Nat'l Science Foundation	4 440	4 479	**5 175**	696	15.50%
Agriculture	2 275	2 309	**1 952**	–357	–15.50%
Commerce	1 073	1 138	**1 152**	14	1.20%
Interior	647	676	**618**	–59	–8.70%
Transportation	767	820	**902**	81	9.90%
Environ. Protection Agency	557	548	**541**	–7	–1.30%
Veterans Affairs	819	891	**884**	–7	–0.80%
Education	327	321	**324**	3	0.90%
Homeland Security	996	992	**1 033**	41	4.10%
All Other	786	819	**821**	2	0.20%
Total R&D	141 933	142 441	**147 361**	4 920	3.50%

Source: AAAS analysis of R&D in the FY 2009 budget (Part 2 of 2).

Box 6.4 The federal R&D budget

For 2009, the federal budget in R&D will total $147.4 billion. With an allocation of $10.5 billion, the DoE is a clear winner among R&D agencies. DoE Science will stand to gain a 21 per cent proposed increase to $4.4 billion for its total budget in order to support the target of doubling its budget between 2006 and 2016. In 2009, most of the DoE's science programmes will receive substantial increases and hit historic highs. After being deleted in 2008, the US contribution to the multinational International Thermonuclear Experimental Reactor project on fusion research will total $493 million, including a $215 million ITER contribution for 2009.

Regarding DOE's energy R&D, it will total $2.4 billion. The most significant increase will be for fossil fuel R&D which will climb to $625 million and will be essentially focused on coal ($624 million): the FutureGen programme (a project to develop a near-emissions-free, coal-fired electricity and hydrogen production plant) will see its budget more than double to $156 million. Funding for the Clean Coal Power Initiative programme to develop cleaner coal-based power plants will total $85 million and carbon sequestration research will total $149 million.

At the same time, some change of priorities has occurred: budgets for hydrogen technology and solar energy are decreasing on the one hand, but biomass R&D continues to grow. In addition, the geothermal R&D programme that DoE had proposed for elimination a few years ago will instead get a boost. Wind energy R&D will receive $53 million, while the former energy conservation accounts will reach $348 million.

Regarding nuclear energy R&D, the budget is some $630 million, partially from the transfer of programmes from other DoE accounts and partially from a proposed increase in advanced fuel cycle R&D because of its key role in the administration's signature Global Nuclear Energy Partnership to promote spent nuclear fuel recycling. Funding for other energy programmes will increase more moderately.

The situation of the Environmental Protection Agency is quite different. This agency's R&D portfolio of $541 million will entail a $7 million cut. In 2009, EPA's R&D funding could fall to the lowest level in more than two decades (since 1985) in real terms, a situation that could be analysed as reflecting poor interest in environment issues at the federal level. Nevertheless, many other agencies have environmental responsibilities related to research, resource stewardship and economic management of the environment. For this reason, EPA is a relatively small funding source for environmental R&D, accounting for only 4 per cent of total federal support. For instance, the Climate Change Science Programme funding will climb over $2 billion, principally supported by environmental science programmes at NSF and DoE Science and by the restructuring of NASA spending to boost spending on earth sciences.

combined. The government pursues a broad range of strategies to reduce polluting emissions. The electric power sector constitutes a privileged sector for developing environmental policies: it is by far the most polluting sector in the United States and compared to other countries, its environmental performance is very poor. It is interesting to note how states are becoming actively involved in environment issues in their power sectors while the federal level is still prudent. State initiatives address GHG emissions by incentives towards less emitting sources of energy as well as energy efficiency among others. The decision of the Supreme Court in April 2007 to confirm the role of the Environmental Protection Agency (EPA) in the regulation of the national emissions level could change the future significantly. The Supreme Court ruled that carbon dioxide is a pollutant under the Clean Air Act and that the EPA has the authority to regulate carbon dioxide emissions from automobiles and other vehicles. Businesses, recognising the change and the market opportunities it signals, have started investing heavily in green technologies and environmental strategies. More and more analysts agree that some significant new initiatives could be taken at the federal level, even if the political process is complex for getting such federal regulation.

3.2 States have the leadership in environmental policies

In the absence of comprehensive federal policy, states are leading the way in addressing climate change issues. Through their authority over areas affecting the environment, such as land use, transportation, utilities, and taxation, states are creating their own programmes and policies that lessen their contribution to climate change.

3.2.1 GHG emissions voluntary programmes

It is interesting to remember that in the early 1990s the United States launched the first market-based initiative in the field of the environment, with the creation of a market for SO_2 emissions permits to address the problem of acid rain. This initiative inspired a lot of other countries to follow suit, notably the European Union that started implementing its own climate change policies.

Regarding GHG emissions, even if the federal legislation does not address the question of capping their level, some individual or collective initiatives have been launched. Many states are adopting policies that successfully reduce polluting emissions. Admittedly these measures have proven controversial at the federal level, such as renewable portfolio standards and mandatory GHG reporting, but they have often been implemented with little dissent in some states. States are taking a range of approaches, from cross-cutting programmes to those more narrowly focused on issues such as energy, air pollution, agriculture, transportation, natural resources and education. While some state programmes are expressly designed to confront the challenge of global warming, others are designed to achieve different policy goals but have the additional effect of reducing GHG emissions. In 2008, twenty-eight states and Puerto Rico developed or started to develop strategies or action plans to reduce net GHG emissions. Several states have set numeric goals for reducing emissions to mitigate climate change. Thirty-nine states and Puerto Rico have completed GHG inventories of their total emissions. To be more precise:

- California, Washington, Oregon, Arizona and New Mexico have launched the Western Regional Climate Action Initiative, which has committed to a multisector cap-and-trade programme. Policy proposals were expected for 2008.
- In December 2005, seven states launched the Regional Greenhouse Gas Initiative. This project has created a CO_2 cap-and-trade for the power sector including all power plants of at least 25 MW of capacity. The programme is planned to begin in 2009 and in a first step it aims at achieving a 10 per cent reduction by 2019.
- Some other states have decided to act independently, such as Washington which has voted a law to reduce its GHG emissions to 1990 levels by 2020 and 25 per cent below 1990 levels in 2035.

3.2.2 Renewable portfolio standard

The lack of consensus on a federal Renewable Portfolio Standard proposed to be enacted into the Energy Security and Independence Act of 2007 has not deterred some states from implementing such a programme at their own level or with other states. The widely different availability and therefore cost of renewable resources across regions in the United States explains the opposition of states with poor renewable power plants or abundant resources of coal. Nevertheless, some major initiatives have been launched such as one gathering twenty-eight states for implementing mandatory or voluntary renewable power mandates (or Renewable Portfolio Standards). Many of these programmes require a portion of the electricity supply to come from a renewable power source such as wind, sunlight, geothermal heat, biomass and some forms of water power. Most of these states are now requiring 15–20 per cent of electricity be supplied from renewables by 2020. This ambitious goal is supported by incentives such as purchase obligation or subsidies given the high cost of these technologies.

Despite growing support, renewable energies are not really developed in the United States. For example in 2008, wind power provides only 1 per cent of United States energy. It could provide 20 per cent of domestic energy needs by 2020. To successfully address energy security and environmental issues, the nation needs to pursue a portfolio of energy options. Texas is the leader in wind power development, followed by California: it hosts the largest wind farm in the world with a 720 MW capacity. It results from its Competitive Renewable Energy Zone Initiative (CREZI) which increases Texas's Renewable Portfolio Standard (RPS) requirement to 5.8 GW by 2015 (with a target of 10 GW by 2025). It also requires the construction of all the transmission capacity needed to allow the renewable portfolio standards goals to be met. The purpose of the CREZI is to identify the best places in the state for renewable energy development (particularly wind generation), and to build new transmission lines to deliver renewable energy to customers. Texas was able to take the lead for several reasons: it is quite isolated from the other parts of the American electricity system and it benefits from a high potential of wind power. Nevertheless, some problems encountered in dispatching the wind power could influence the future path of this ambitious policy.

3.2.3 Energy efficiency

At the same time, the rising cost of power generation due either to price hikes for primary energy from thermal power plants or to the poor competitiveness of renewable energy sources, is encouraging some states to develop ambitious programmes in the field of energy efficiency to compensate the final consumer with the lowest consumption. As federal legislation imposes efficiency standards for appliances, equipment and buildings, more and more states are setting standards exceeding the federal ones (e.g. New Jersey, Ohio, Kentucky). Other states offer direct

Box 6.5 California and GHG emissions

California has enacted the most progressive renewable energy and demand response initiatives in the country with a Renewable Portfolio Standard of 20 per cent by 2010 (which will most likely rise to 33.3 per cent by 2020), a $2.9 billion solar initiative, and an accelerating demand response regime. The state is about to go further.

- *Situation*: The combustion of fossil fuels accounts for 88 per cent of California's GHG pollution. According to the Californian commission, transportation is responsible for 41 per cent of GHG emissions; state electricity generation for 10 per cent and industrial facilities 23 per cent (over 40 per cent of petroleum refineries). The State of California is the twelfth largest source of global warming pollution in the world.
- *Law*: California passed its climate change law in August 2006 and is now implementing the regulatory framework to meet a 2012 deadline to cut by 25 per cent the 1990 levels in greenhouse gases by 2020. This reduction will rise to 80 per cent by 2050. The law is mandatory, not following the path of the voluntary markets that the federal administration is touting. California is so serious about greenhouse gas reductions that the measure was passed in an election year, was bipartisan and contributed to the re-election of Governor Schwarzenegger.
- *Cap and trade system*: California will be implementing a mandatory cap and trade system although it is not written into the law. How will the allowances be allocated? Will they be facility-specific or industry-specific? How will the 1990 baseline be determined? How many carbon credits will be auctioned? There is also going to be some percentage of credits buying in the Kyoto international markets for California's emissions compliance.
- *Transportation sector*: Clean car regulations have resulted in litigation with automobile manufacturers with the unfortunate position of the Californian government suing the US EPA on a waiver under the Clean Air Act which will allow California to regulate tailpipe emissions. The low carbon fuel standards are an attempt to reduce California's 98 per cent oil dependency.
- The Western States Climate Initiative was created to broaden the Californian climate change efforts into a regional solution.

funding for energy efficiency projects in public or private buildings (Montana, Oklahoma and Alabama). And others take initiatives in all these categories of incentives (California, Oregon).

A growing number of major companies are also making significant efforts to address climate change. These efforts include setting GHG reduction targets, improving energy efficiency, investing in the development of clean and renewable energy technologies, increasing the use and production of renewable energy, improving waste management, investing in carbon sequestration, participating in emissions trading, and developing energy-saving products. Companies are acting in the absence of mandatory requirements on the supply side. Deregulation has also brought about a fascinating creation of new business models and dynamic corporate thinking. Competitive utility retail businesses are both the culmination of the promise of high-technology and a new opportunity.

However, the legislation is beginning to impose more strict regulation on the demand side: the Energy Security and Independence Act of 2007 focuses on electric power demand and the most significant provisions lay out standards and efficiency requirements for lighting, appliances and buildings in terms of the potential to reduce electricity consumption and carbon emissions. These provisions reflect one of the major features of the current energy and environmental policy of the United States: its confidence in technological progress to solve the environmental dilemma. The federal support for major research and development programmes in the field of hydrogen, capture and sequestration of CO_2 is directly inspired by this 'philosophy' but a considerable number of individual states are currently launching or plan to launch their own environmental programmes either alone or collectively.

4 Conclusion

What will the American vision and role be in building the global energy future? The United States has refused to ratify the Kyoto Protocol. Nevertheless, more and more people, states and municipalities are beginning to realise that climate change is a real challenge and that something has to be done. At the same time, the United States is worrying about its growing dependency on imports and the question of security of supply is becoming a major concern.

A new combination of the federal and state level decisions is emerging that could help the United States to face the new challenges created by a world in which energy will be rare and more expensive than in the past.

At the federal level, most decisions address the question of security of supply: new arbitrages in the allocation of R&D budgets have been done in order to develop clean coal technologies that could help to reduce both emissions and dependency from imports; fiscal as well as financial incentives to new nuclear plants would help to maintain the role of this technology in the future electricity mix. The foreign policy of the United States also supports this major concern of security of supply, even if some decisions could face more or less violent criticisms.

At the state level as well as at the level of citizens, new trends are emerging: the decrease of demand for gasoline which could reflect a structural change in the consumption of 'transportation'; the implementation of ambitious programmes of renewable energy sources; measures to promote energy conservation; use of alternative fuels; and a renewed interest in programmes of demand-side management.

That the American way of life is not negotiable has been one of the arguments used for refusing the ratification of Kyoto Protocol. But environmental issues, particularly concerns about climate change, have now taken on a much higher profile in the United States. As a result the United States energy policy is at a turning point: the federal level focuses on the security of supply issues but citizens and with them politicians at the state level are worrying about the environment. This apparent opposition could be the beginning of a new 'energy deal' in the United States: 'We will harness the sun and the winds and the soil to fuel our cars and run our factories (President Obama, Inaugural Address, January 2009).

Notes

1. Coal use has gone up 211 per cent and electricity from coal has increased 179 per cent.
2. By comparison, China is the second largest consumer of oil at 7.2 Mb/day and Japan the third with 5.2 Mb/day.
3. Order 888 mandated the unbundling of electrical services and the separation of marketing functions for these newly disaggregated services, required utilities to provide open access to their energy rate schedules and gave existing utilities which may have made investments based on older regulations the right to recover their stranded costs. Order 889 set standards regarding information that utilities must make available to the marketplace and established OASIS, a bulletin board system for sharing this information.
4. Negawatt power is a technique that works by investing to reduce electricity demand instead of investing to increase electricity generation capacity.
5. For an economic analysis of the Californian crisis, and the critical issue of market power in power markets, see Borenstein S., Bushnell J. & Wolak F. (2000).

6. The Climate Stewardship Act of 2003 (S.139) was introduced in January 2003 by Senators Joseph I. Lieberman (D-Connecticut) and John McCain (R-Arizona).

References

Borenstein, S., Bushnell, J. and Wolak, F. (2000) *Diagnosing Market Power in California's Deregulated Wholesale Electricity Market*, POWER, Working Paper PWP-064.

Chevalier, J.-M. (2008), *Les 100 mots de l'énergie*. Paris: Presses Universitaires de France, Collection Que sais-je?

Energy Information Agency, Department of Energy (DoE) (2008) *Annual Energy Outlook 2008 with Projections to 2030*. Report No. DOE/EIA-0383(2008).

Joskow, P. and Schmalensee, R. (1983) *Markets for Power*. Cambridge, MA: MIT Press.

Kahn, A. (1988) *The Economics of Regulation: Principles and Institutions*. Cambridge, MA: MIT Press.

Platts Nuclear Publications (2007–9).

Yergin, D. (2008) 'Oil at the Break Point', Special Report, Testimony by Daniel Yergin, Chairman of Cambridge Energy Research Associates, before the US Congress Joint Economic Committee in Washington, DC, 25 June 2008.

Zaleski, C. P. (2006) 'The Future of Nuclear Power in France, the EU and the World for the Next Quarter-Century', *Perspectives on Energy*.

7

Climate Change, Security of Supply and Competitiveness: Does Europe Have the Means to Implement its Ambitious Energy Vision?

Jan Horst Keppler

1 Energy policy, a cornerstone of European integration, still needs to come to terms with the new realities

Energy questions have always played an important part in shaping the identity of modern Europe. Right from the start, the first common institution of the original six countries of the European Union was the European Coal and Steel Community (ECSC) in 1951. It was followed in 1957 by the European Economic Community (the EEC) and the European Atomic Energy Community (EAEC or EURATOM). The founders of modern Europe were aware of the strategic character of energy security. The European Union would be an energy union or it would be nothing. Recently, the French presidency of the European Union that began on 1 July 2008 chose energy as a priority topic beside climate, immigration, defence, the reform of the common agricultural policy (CAP), the Union for the Mediterranean and Social Europe. Its key task in the energy area is in passing the vastly ambitious second legislative package on energy and climate that proposes to shape European energy policy up to 2020 and beyond.

This brief institutional background is helpful in order to characterise Europe's current situation in the field of energy. After years of slow drift, the question of pooling energy stakes is again at the heart of European policy-making. For example: (1) the liberalisation of grid-bound electricity and gas markets is today the key ambition of the European Commission as far as industrial policy is concerned; (2) the bold policy to introduce the first major market for CO_2 emissions – a global first – and to reduce emissions by 20 per cent below their 1990 level by 2020 (by 30 per cent if other countries follow suit) is an area in which Europe can pride itself on genuine global leadership; (3) managing the relations with oil- and gas-supplying countries to ensure energy supply security, whether they be Russia, the republics of central Asia, Iran or the countries of the southern rim of the Mediterranean. These constitute today the central aspects of European foreign policy.

Of course, in none of these three fields does the Union and its principal agent, the European Commission, proceed without difficulty. However, in a phase when the process of European integration is at a standstill, the debate about energy issues is part of the larger debate about the nature and the destiny of the European Union. Managed well, it could become a catalyst for a renewal of the European identity, particularly since, in the climate area, policy-makers proceed on the basis of a large popular consensus, which is more than can be said for many other policy areas. In the event of failing to advance the energy and climate agenda, however, the questions about the purpose of the European Union would become increasingly pressing.

Here the similarities with the situation in the 1950s end. The differences are also marked. In the years after the Second World War, the key question was how to rebuild industry, in particular heavy industry of strategic value, whose installations had been largely destroyed. Government involvement, subsidising renascent European industries and market protection were the watchwords of the day. While not in keeping with economic orthodoxy even at the time, such a productivist industrial policy could be amply justified for two reasons. First, in the absence of a functioning market, governments had to take the lead and become involved. Second, even if any inefficiencies arose from this involvement, they would be amply outweighed by the political pay-off of binding together the countries of a continent that had been ravaged by war twice in a generation by pooling their strategic energy assets.

Today the situation is different. Partly due to implementing a shared energy vision, European countries have enjoyed peaceful coexistence. The difficulty is rather to adapt the historical vision to the realities of today and to overcome conditioned reflexes that are no longer relevant and may even create more harm than good when aiming to create the energy Europe of tomorrow. Today, the European Union is not only a major economic power with leading world companies in each segment of the energy market, but it is also far more closely integrated with the rest of the world than it was in 1951. Moreover, the importance of the domestic resource, coal, has considerably decreased in favour of oil and gas, largely imported resources. To face up to these new realities, Europe cannot allow itself to remain steeped in an outdated rhetoric. In the

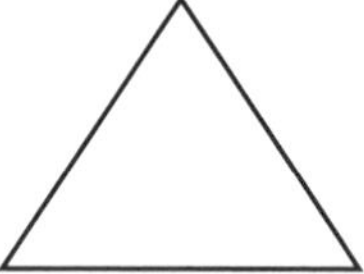

Figure 7.1 The triangle of European energy decision-making

modern energy world, there is no place for visions of manifest destiny. The current situation does not require introspection with respect to mutual subsidisation, but the acceptance of a global interdependence (on the energy as well as on the climate level) and the creation of structures which support the reactivity, competitiveness and sustainability of European energy industries.

Such a fundamental commitment towards open and competitive energy markets (while taking account of technical constraints, for instance in the electricity sector), by no means impedes the safeguarding and the development of common goods such as the environment or energy security. The essence is to understand that these important policy objectives must be pursued with the help of instruments that are compatible with the logic of the market (tax and price policy, emissions trading, general targets) rather than with instruments that go against this logic (administrative oversight, selective subsidy, choosing technologies, setting micro-targets).

The commitment to an old-style industrial policy in the energy sector served Europe well for many years. Now, however, it is threatening to hold it back. The difficulty of abandoning the mindset of the past shows up, in particular, in the current inability of European policy-makers to define and decide on the organisation of trade-offs between three key policy objectives, which are energy security, industrial competitiveness and environmental protection (see Figure 7.1). The dithering of the policy-makers in the European Commission and national governments adds to the objective difficulties Europe is facing. Reducing greenhouse gas emissions, limiting subsidies, decreasing import dependence, phasing out nuclear power, augmenting the use of renewable energies, liberalising energy markets, increasing economic competitiveness ... the wish-list of energy policy objectives is very long indeed and not every addition is being carefully considered with respect to its consequences.

The contradictions in Europe's triple policy orientation show up in very concrete terms. Consider, for instance, the share of gas in European energy consumption. Favoured over coal by environmental considerations and over nuclear on cost, natural gas very well satisfies objectives 2 (environmental objectives) and 3 (competitiveness). Unsurprisingly, its share in total primary energy supply is expected to rise from 23 per cent in 2004 to 30 per cent in 2030 (IEA 2006). Increased natural gas consumption, however, means increased import dependency and thus contradicts objective 1 (security of supply). The inability to define lasting trade-offs between the various objectives implies continuing drift.

Matters are made worse by an institutional vacuum in the energy field. The Directorate General responsible for energy, the DG TREN, has a difficult time making its voice heard amidst the politically stronger directorates responsible respectively for competition policy, environmental policy and external relations. In 2007, the President of the European Commission, José Manuel Barroso, even created his own high-level Advisory Group on Energy and Climate Change. The fact that energy is now a politically attractive issue should be a good thing. Unfortunately, the attention lavished on energy questions is only haphazardly related to effective policy-making.

Crucially, European efforts to forge a proper energy policy are hampered by the lack of an internal consensus about the nature of the trade-offs between competing policy objectives (see Figure 7.1). The matter is one of style as much as of substance. In the absence of a commitment to price instruments such as energy taxes, European policy-makers hope that 'energy efficiency improvements' might let them have their energy cake and eat it too – oblivious to the fact that energy efficiency improvements depend largely on price. Higher prices and taxes on energy instead would not only promote environmental quality and security of supply but would also allow, via a budget neutral recycling of tax revenues, the cutting of payroll taxes and social security contributions, the bane of European industrial competitiveness.[2]

With many European companies operating in highly competitive global markets and the European economy lagging in dynamism, any increase in energy prices will have to be argued for very carefully. Without higher energy prices, however, substantially reducing demand increases or lowering energy intensity will remain elusive. The creation of the

European Trading Scheme (ETS) that prices CO_2 emissions and thus implicitly raises the price of carbon-emitting hydrocarbons such as oil and gas, is the single most promising policy measure of recent years to improve not only environmental sustainability but also European energy supply security. In order to improve on this positive contribution, transport and other sectors with non-point sources with their massive oil consumption should be included as quickly as possible into the ETS.

2 Strengths and weaknesses of the European energy situation

Having outlined the general context of European energy policy-making, it is now time to look at the European energy situation in more detail. The main weakness of the European energy situation is, of course, its relative lack of domestic hydrocarbon resources. Oil and gas need to be imported at ever-higher prices for 80 per cent and 60 per cent of consumption respectively. These percentages are set to increase further, due to the rapid decline of oil and gas fields in the North Sea. Of course, Europe has plenty of coal but given its ambitious target of reducing carbon emissions by at least 20 per cent by 2020, coal is not a very attractive option at least as long as carbon capture and storage (CCS) remains too expensive to be implemented on an industrial scale.

Europe also possesses considerable strengths in the energy sector. The overall energy efficiency of its economy is higher than Japan's but considerably lower than that of the US. It has world-class competitors in virtually every segment of the energy market whether it be in oil and gas (BP, Shell, Total), electricity production (EDF, E.ON), nuclear reactor construction (Areva), turbine construction (Alstom, Siemens) or renewables (Vestas). Its physical and institutional infrastructure is of very high quality.

And yet, despite the best efforts to promote renewable energies and induce structural changes, Europe's energy situation in the future is in danger of remaining quite similar to that of today. Energy consumption and, in particular, power generation, will continue to depend heavily on fossil fuels. In 2004, coal constituted, according to the most recent figures of the International Energy Agency, 18 per cent of total primary energy consumption, oil 37 per cent, and gas 24 per cent. The balance was provided by nuclear power and renewable energies with approximately 20 per cent (see Figure 7.2). The high growth rate of the use of renewable energy in the IEA's 'Reference Scenario' (the prolongation of current trends) of nearly 5 per cent per year (compared with the growth

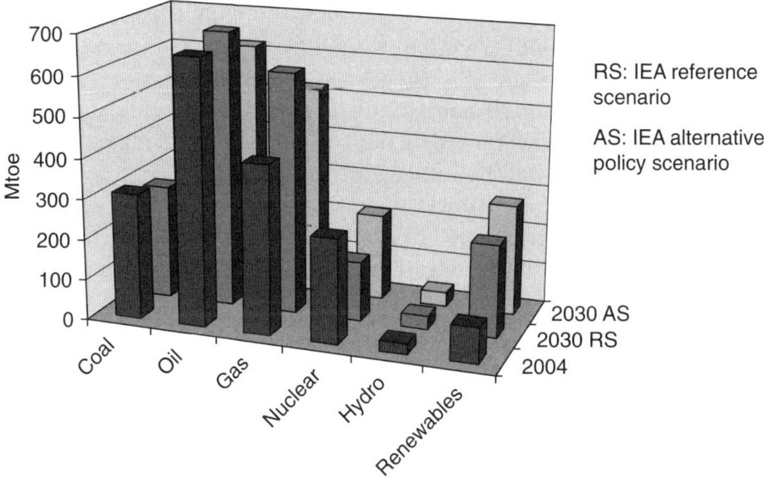

Figure 7.2 The European primary energy supply, 2004 and 2030
Source: Author's calculations based on available data from WEO, IEA (2007).

rate of the total primary energy supply of 0.5 per cent and less than 0.2 per cent in the more proactive alternative policy scenario) will not bring a dramatic change to this situation since it starts from a low base.

Because of the inelasticity of energy demand in the transport sector, any impulses for structural change in the European energy sector will need to come from the power sector. Coal and nuclear power here each represent 31 per cent of total production, while gas provides 19 per cent, hydro 10 per cent, renewables 5 per cent and oil 4 per cent. The IEA expects shares in gas and renewables to grow quickly (at 3 and 6 per cent respectively in a market whose growth is 1 per cent per year) to reach 32 and 19 per cent respectively of total electricity production in 2030. This is supposed to go hand-in-hand with the decline of the shares of coal and nuclear power. On the latter point, the IEA forecasts are probably too pessimistic. In 2030, Europe will use more, not less, nuclear power than today. However, the IEA forecasts need to reflect official government positions – such as the announced phase-outs in Germany and Sweden – even when the realities underlying the debate have moved on. Its share in the production of electricity could even increase relatively quickly if the announced renaissance of nuclear power is confirmed by further orders after the construction of new EPR reactors in Finland and France.

There are three key reasons why European energy prospects do not differ appreciably from the current situation:

1. The inelasticity of oil demand for private transport; technical progress is lacking, although, spurred by high oil prices, several major projects trying to develop electric vehicles are underway;
2. The continued attractiveness of natural gas as a fuel of choice of private investors in Europe's liberalised power markets (see below);
3. The incapacity of the European decision-makers to agree on a coherent energy policy (see above).

However, it would be unjust to characterise the European energy situation as one of complete stagnation. The ambitious policy objectives announced in the second legislative package on energy and climate may yet lead to structural shifts which the IEA forecast has not captured. These would, however, require a long-term commitment to permanently higher energy, carbon and electricity prices which European policy-makers are currently still too timid to embrace.

2.1 Sector focus 1: the gas market

For the time being, natural gas is the adjustment variable when comparing the reference scenario and the alternative policy scenario. Partly because of this, gas is also the fuel that has attracted most attention in recent debates about European energy security. This is due to its rapid rise in the European energy mix as an economically and environmentally attractive fuel. Gas-fired combined cycle turbines (CCGTs) have become the technology of choice for private investors faced with the uncertainties of Europe's liberalising power markets. With low investment costs and short lead-times, CCGTs allow investors to react quickly to changing market conditions. In addition, gas is frequently the marginal fuel in the market and thus sets the electricity price. This means private investors are largely shielded from the negative impacts of high or volatile gas prices. If the gas price goes up so does the electricity price; the pay-off for the investor remains the same, by and large. Add to this that gas emits about half as much carbon as coal per MWh of electricity and the advantages outweigh the disadvantages.

Together with a mature transport infrastructure and a good security of supply record, the attractiveness of natural gas has contributed to an *increasing* intensity of gas consumption since the 1970s. Europeans today use more gas per unit of GDP than thirty years ago.[3] This implies some risks. Further price increases (highly likely in particular once Russia begins exporting gas to Asia), or physical supply interruptions (very unlikely, although the two-day interruption of gas deliveries to Ukraine unsettled European commentators) may thus have larger impacts on the European economy than otherwise. Nevertheless, Europe is likely to consume in 2030 not only more nuclear but also more gas than today. Europe, while possessing only limited reserves of its own, is also favoured by geography in that it is within 'pipeline distance' (5,000 km or less) of two-thirds of the global gas reserves (see Figure 7.3). This is no negligible detail. It means Europe can depend on a diversified set of suppliers by pipeline in addition to the growing quantity of LNG deliveries.

Total gas imports of the 27 member countries of the European Union were 307 billion cubic metres (bcm) in 2007, roughly 60 per cent of total consumption. Main supplier countries were Russia (40 per cent of total imports), Norway (28 per cent) and Algeria (11 per cent). The most dynamic section of the gas market is the global LNG trade which already constitutes 15 per cent of imports and is rising fast. It is estimated that European LNG imports will rise by 7.5 per cent annually, compared to 5.1 per cent for imports through pipelines and 2.1 per cent for the growth of total demand. Global LNG trade will be fuelled by Qatar's massive 'North Field' of an estimated 900 trillion cubic metres, which constitutes by itself 14 per cent of proven global gas reserves.

The most importing trading partner for Europe in the gas market, however, remains Russia which supplies a quarter of total gas consumption. While Russia has been a reliable supplier of hydrocarbons for decades, two recent episodes made the headlines. In the winter of 2005/6, the dispute between Ukraine and Russia over gas tariffs slightly reduced European supplies for several days. The event was unsettling on a symbolic rather than on an economic level, leading only to a minor shortfall of 100 million tonnes, which corresponds to a difference in demand due to a temperature change of 2 degrees Celsius on a single day (Ladoucette 2006: 4). A similar dispute with Byelorussia concerning the Drushba oil pipeline one year later did nothing to improve the situation.

Other issues weighing on energy relations are the fast-developing international context, including the rise of LNG, the proposed coordination between Russia and other supplier nations, the uncertainties of Russia's potential to increase exports in the face of rising domestic demand, as well as the potential of the Asian market for Russian exports.[4] Despite a number of misunderstandings and a bilateral summit without major breakthroughs in May 2007, both Europe and Russia are trying to put their energy relations on a sound basis. Three new working groups – on Energy Strategies, Forecasts and Scenarios, (b) Market Developments and

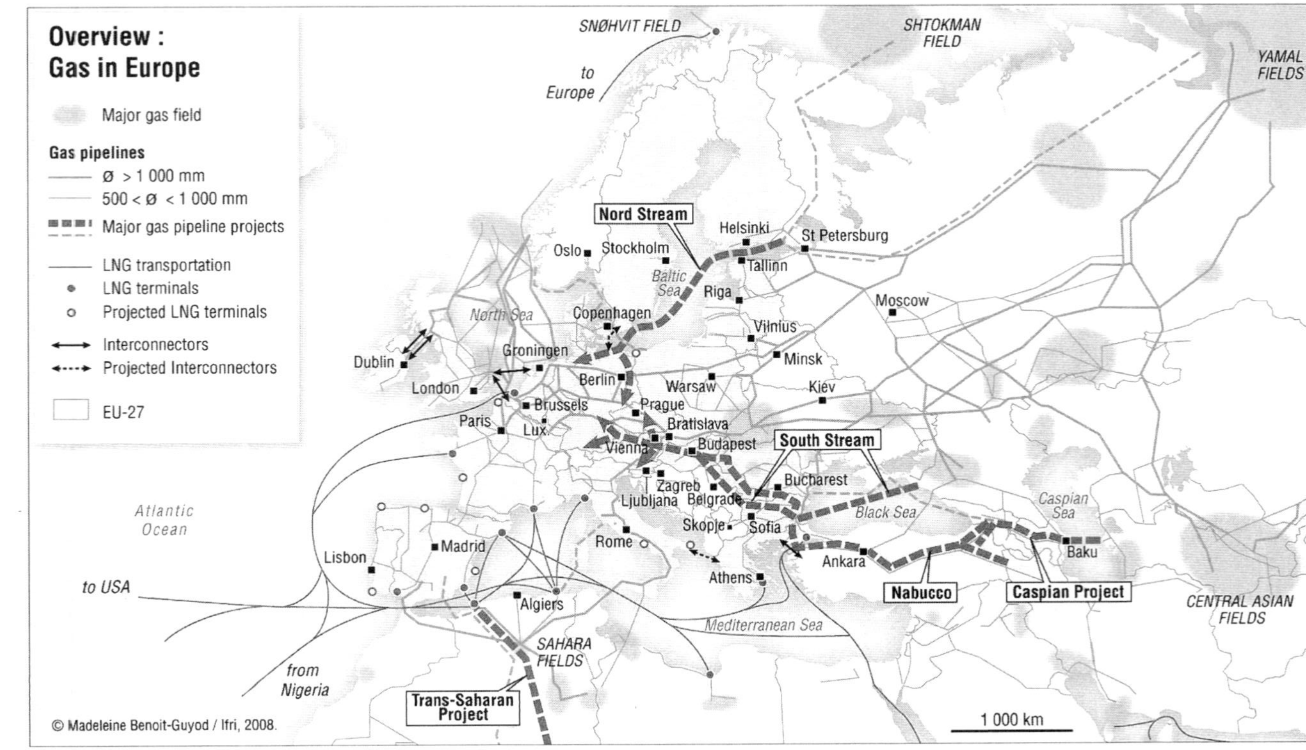

Figure 7.3 Gas in Europe
Source: IFRI (2008).

(c) Energy Efficiency – are intended to foster a better understanding of the respective points of view. A new framework agreement for cooperation was negotiated in November 2008.

In evaluating Europe's dependence on Russian gas imports, one should not overlook the fact that the 121 bcm that Russia exported in 2007 to Europe constituted the bulk of Russia's total exports of 148 bcm. In addition, it constitutes by far the most profitable part of Russia's huge annual production of 607 bcm fuelled by a domestic consumption subsidised with prices of around one quarter of world prices. Gas exports to Europe are estimated to constitute 70 per cent of Gazprom's revenues (Finon and Locatelli 2006: 8). Nowhere is the old adage that dependence implies interdependence more true than in the gas trade between Russia and the European Union.

The gas market is also part of the backdrop for the new French initiative to create a Union for the Mediterranean. Not only are Algeria and Libya important suppliers (with Egypt hoping to join them soon), Mediterranean countries such as Turkey, Greece or Croatia are also potential hosts for the pipelines proposed to bring gas from either southern Russia, central Asia, the Caspian Sea, Iran, Iraq or even Qatar. These are politically sensitive issues. This sensitivity is brought out in stark relief by the competing pipeline projects of Nabucco (sponsored by the European Commission) and South Stream (sponsored by Russia), both designed to bring gas through south-eastern Europe to Central and Western Europe (see Figure 7.3). Since in the end only a single pipeline could profitably carry all the gas required, there is a chance that reason will prevail – perhaps in the form of a joint project with two trunks. The episode, however, shows that for Europe gas is not yet a commodity like any other.

2.2 Sector focus 2: the power market

Electricity is the largest energy-consuming sector of the European economy and the one where gas demand is growing the fastest. Given the historic inelasticity of energy demand in the transport sector, any impulses for significant structural change in Europe's energy sector will have to come from the power generation sector. As we noted above, coal and nuclear each represent 31 per cent of total electricity generation, gas 19 per cent, hydro 10 per cent, renewable energies 5 per cent and oil 4 per cent. Expectations are that gas and renewable energies will grow fast (at 3 and 6 per cent per year respectively in a market that grows at 1 per cent per year) to reach 32 and 19 per cent respectively of total electricity generation in 2030 in the IEA's 'business as usual' reference scenario. This will go hand-in-hand with declining shares for coal and nuclear. Higher gas and electricity prices (not least due to pricing for CO_2 emissions permits) combined with improved efficiency of power consumption might also lead to lower growth in the electricity market.

Coal-fired power generation, in particular *new* coal plants, will be progressively priced out of the market by higher prices for CO_2 emissions. As has been said, the share of nuclear energy might still increase beyond the forecasts. In the absence of confirmation of a nuclear revival, however, gas remains the most likely option for capacity increases, assisted by the continuing drive of the European Commission's Directorate General for Competition Policy (DG COMP) for ever more complete liberalisation of the electricity market. While the liberalisation of electricity markets is driven by the laudable motive of improving the efficiency and competitiveness of European power markets, it has also been the source of increased price volatility and much regulatory uncertainty.[5] In some cases, such as on the issue of long-term contracts, the Commission seems driven more by the theoretical coherence of its approach rather than by the realities of a sector in which producers demand nothing more than some visibility for long-term investments.

Europe's power sector is thus under tension. It pits industry against the Commission and governments, producers against consumers and liberalisers against advocates of continued government involvement, while the nascent regulatory agencies flex their muscles without having yet established the sort of predictable reflexes that would allow them to provide guidance to the market. This results in increased uncertainty and a lack of investment. Europe has experienced an increasing number of blackouts: in France (1999), London (2003), Denmark and Sweden (2003), Italy (2003), Greece (2004), Spain (2004) Germany (2004) and Western Europe (2006) (Ladoucette 2006: 5). While in some instances technical issues played a role, the fact is that electricity demand increasingly outstrips supply in European markets. Overall, the average capacity margin in the Europe was 4.8 per cent in 2005, down from an already low 5.8 per cent in 2004 (Capgemini 2006: 3).[6] Unsurprisingly, electricity prices have increased rapidly in recent years (by about one-third each year since 2002), a tendency that was reinforced in January 2005 by the introduction of the European system for CO_2 emissions trading, the ETS (see Figure 7.4).

There are other factors that disturb the (imperfectly but increasingly) integrated European electricity market. German investments of more than 20 GW in wind energy and chronic undercapacity in Italy have led to massive North–South flows of electricity since 2004 without notable impacts on the security of energy supplies (see Keppler 2005). The two

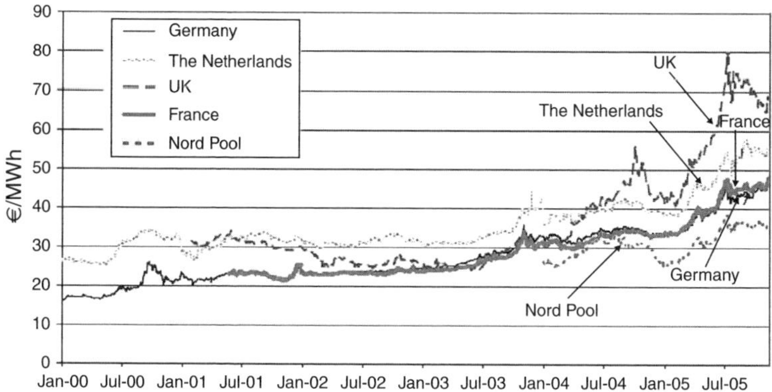

Figure 7.4 Wholesale electricity prices in Europe on the rise since 2000
Source: European Commission (2007).

key issues for the security of European energy supplies in the electricity sector are the question of nuclear energy and the financing of future investments. The question is how Europe will generate the funds for investments of 1,400 billion euros up to 2030 in electricity markets, three-quarters of it in generation capacity (Baseline Scenario of the European Commission's Directorate General for Transport and Energy). There are two reasons why current market conditions incite actors to postpone investments rather than aggressively promoting them. First, price volatility raises the question of the implicit rate of return to investors. With inelastic demand in electricity markets, existing producers have every incentive to create a structural undercapacity. Substantial fixed costs that relate to conditions for network access, risk diversification and the combination of technical and financial know-how required in modern electricity markets constitute barriers to entry for newcomers. In the absence of a competent European regulator, there is little chance that the situation will change rapidly.

The situation is even more complicated for investments dedicated to nuclear energy and this despite the most favourable outlook for nuclear in the past twenty-five years. In a carbon-constrained world, in which the European countries are committed to reach their Kyoto targets (a reduction of 8 per cent below the level of 1990 emissions by 2012 and a 20 per cent reduction by 2020), an increase of coal-fired power generation in the absence of carbon capture and storage is not a viable option. The only real alternative is nuclear power generation with renewable energy playing an important but not decisive second role.[7] After an

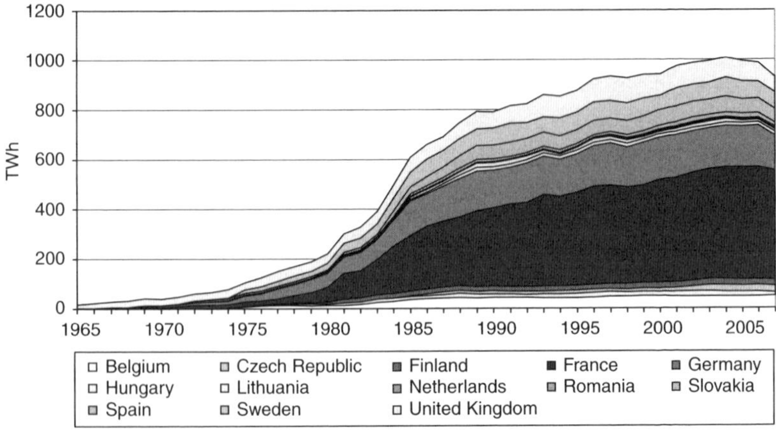

Figure 7.5 European nuclear energy consumption, 1965–2007
Source: Author's calculations based on data available from BP (2008).

uninterrupted increase for more than forty years, nuclear power has begun to decline since 2006 mainly due to Germany's decision to phase out nuclear power and low capacity rates in France (see Figure 7.5). A number of positive factors, however, should encourage the medium-term outlook for nuclear power:

- Rising gas prices have heightened security of supply concerns and made nuclear more competitive.
- Renewable energy sources are still too expensive to provide a credible alternative, while nuclear power saves more than 300 million tonnes of CO_2 (8 per cent of the EU total emissions); the EU's commitment to the Kyoto Protocol limits the upward potential of coal.
- The decision of Finnish TVO to build new reactors in Europe demonstrates economic competitiveness and innovative management of economic risks; France is also building a new European pressurised reactor (EPR) with several European partners.
- The fast-growing Chinese electricity demand is creating demand for nuclear technology exports; while in the United States the 2005 Energy Bill provides for insurance, subsidies (2.5 cents per kWh for new nuclear plants) and waste disposal (Yucca mountain).
- There is greater realism and less emotion in the European debate on nuclear energy; polls indicate that even in countries formally committed to phasing out nuclear such as Germany and Sweden, a majority of the population now has a favourable view of nuclear power.

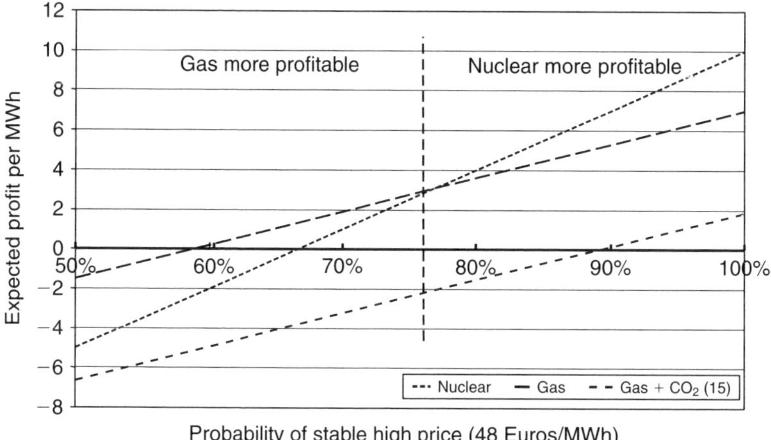

Figure 7.6 The profitability of nuclear decreases with price uncertainty
Source: Author's calculations based on data available from IEA/NEA (2005).

The key factor impeding the construction of new nuclear power plants, however, is the investment risk that comes with the high fixed costs and long lifetimes. This heavily penalises nuclear energy in the eyes of risk-averse private investors who have to face the volatile price environment of liberalised electricity markets. Figure 7.6 captures the essential features of the economic competitiveness of nuclear energy. From left to right, the certainty of stable high prices for electricity decreases (here assumed to be 50 euros per MWh). This implies the increased likelihood of a low-price scenario (here assumed to be 17 euros per MWh equal to the marginal cost of nuclear power). Once the likelihood of a low-price scenario is higher than 25 per cent, gas becomes the fuel of choice for investors even if both technologies are still profitable and even if nuclear power is the cheaper option in terms of levelised average cost (equivalent to a certain high-price scenario).

How can this be? It is crucial to understand that low fixed-cost technologies such as gas, where investment costs are only about 30 per cent of total lifetime costs, come with an option for investors to exit the market if prices fall below a certain threshold. Nuclear power, where investment costs are 70 per cent of total lifetime costs, does not provide that option. If prices fall, investors will be stuck and will continue producing even if they have no hope of ever recovering their money. Risk-averse investors will thus opt for gas in markets with uncertainty about the price for electricity.

However, Figure 7.6 also shows that a stable price for carbon (here assumed to be 15 euros per tonne of CO_2), will make nuclear energy *always* the preferred option for private investors under all circumstances. Of course, such a carbon price would need to be imposed on CO_2 emissions for the whole lifetime of the plant (up to sixty years). The profitability of nuclear energy thus depends on two crucial conditions:

1. The existence of long-term contracts or producer–consumer consortia (the option chosen in Finland and currently explored in France and Belgium) that mitigate or eliminate price risk;
2. Credible long-term expectations about a significant price for carbon emissions.

In addition, the waste disposal issue must be solved at a European level. Collaboration with Russia which possesses the required geographic and geological conditions for waste disposal could be a win–win proposal for both parties.

Given that these conditions will not materialise immediately (even if second period carbon prices are now in their high twenties, there is still little visibility about their level over the lifetime of a nuclear power plant) the European electricity sector will continue to provide a serious challenge to European policy-makers. Wanting to improve security of supply and environmental performance will most likely mean higher prices, in particular if the structural under-investment in new capacity is not addressed. One important point is that policy-makers need to educate the European public that it will not be able to have it all. Inside the triangle of European energy decision-making, trade-offs need to be made, implemented and communicated. In short, Europe needs policy leadership.

2.3 Sector focus 3: carbon markets and the second legislative package on energy and climate

The most innovative initiative in European energy politics was undoubtedly the introduction of the European emissions trading system (EU ETS) for CO_2 emissions in January 2005. While other emissions markets exist, notably the US market for SO_2 emissions, the European initiative is unprecedented in scope and scale. With 2 billion tradable permits worth at current prices 60 billion euros, the EU ETS has managed to create a new asset class that interacts closely with other energy markets, in particular the gas and the electricity market. The EU ETS is also a high-profile global

Box 7.1 Nuclear energy in Europe

Nuclear energy is probably the weakest point in the European Union's energy/climate change policy. Indeed, there is persistent disagreement among EU member states on nuclear power. Policies can change quickly with elections, and the attitudes of different states have changed over time. Today, the majority of European countries are in favour of developing nuclear power.

The governments that do not agree with further nuclear development, or at least are not pro-nuclear, and do not have nuclear power themselves are Austria, Ireland, Luxemburg, Greece, Denmark, Cyprus, Malta and Portugal. Germany and Belgium have nuclear power phase-out laws on the books but the industry and some politicians are seeking a way to keep reactors operating. Sweden also has a moratorium on new reactors, dating from 1980, but the population is in favour of nuclear power. Spain's socialist-led government has a policy to phase out the country's nine nuclear power units, but no law is on the books. We may note that Italy shut all its operating reactors in the 1980s and imposed a moratorium on building new ones, but the new government has declared it wants to reintroduce nuclear power.

It will be important for Europe to speak with one voice, especially if it wishes to influence the world's attitude towards energy and climate change, where the most important countries – the United States, Japan, China, India, South Africa, Russia and South Korea – and many others have a very positive attitude towards the development of nuclear energy.

Indeed, the proponents of nuclear energy point out that nuclear base load electricity occupies the best position on each of the three points of the triangle depicted in Figure 7.1: security of supply, CO_2 emissions and economy. Security of supply will be ensured with the deployment of fast breeder reactors after 2040. For CO_2, nuclear performance is one of the best, and much better than fossil power, including gas, and even better on a life-cycle basis than solar power. As for economic performance, nuclear power costs much less than power from renewable sources, calculated over the lifetime of a plant; nuclear is also cheaper than oil and gas-fired power given the present trend of gas and oil prices (fluctuating but with a general upward trend), and somewhat cheaper than coal-fired power if a minimum value is assigned to CO_2 emissions (€20 per tonne; this would increase the cost of electricity generated from coal by about €15 per MWh) or if carbon capture and sequestration is developed and implemented.

The capital cost per kWe net of a modern coal-fired plant, including CCS, would likely be the same or higher than the capital cost of nuclear plant kWe. That makes the issues of high initial capital cost and the likely increase in the cost of building materials similar for both technologies. The cost of the fuel cycle over the lifetime of a power plant will thus clearly favour nuclear. As far as private investors are concerned, the decision between gas and nuclear will ultimately depend on the stability of electricity prices and the perceived prospects for gas supplies and prices.

The opponents of nuclear energy often mention three main issues: the unsolved waste problem, the large potential consequences of a nuclear accident, and the influence of civil nuclear energy deployment on the proliferation of nuclear weapons. For space reasons, it is impossible to discuss these important issues in this chapter, but they must be solved correctly if nuclear energy is to play a major role in the future.

However, these issues are not country-specific issues; they are European or rather even global issues. It is therefore likely that the world, whose major countries, including some in Europe, are going to develop nuclear energy, will have to resolve these issues independently of the viewpoints of some European countries. An additional issue is the diversity of Europe's regulatory structure for nuclear installations. Some consolidation between European countries developing nuclear energy would certainly help in creating a more level playing field for Europe's nuclear industry. Overall, if Europe is serious about reducing its CO_2 emissions and increasing its security of supply, it will need to develop nuclear energy more dynamically.

C. Pierre Zaleski (CGEMP).

initiative since it is Europe's principal instrument for achieving its Kyoto targets.[8] It is fair to say that Europe has been the principal global champion of the Kyoto Protocol, the set of greenhouse gas emissions reduction objectives solemnly pledged at the third conference of the parties held in Kyoto in 1997. Without European leadership and political pressure the Kyoto Protocol would not have been able to assemble the necessary qualified majority to come into force.

Action against climate change has been the one area in which Europe has projected a cohesive and decisive policy initiative in the international arena. It is also the one major policy issue in which European leaders and their citizens see eye-to-eye. This consensus, rather precious in times of Euro-scepticism and a general wariness of new policy initiatives, is driven by two factors, one explicit and the other implicit.

The explicit factor is that the European public is seriously concerned about climate change. The Stern Review, the policy-oriented compendium on all things climate-related, compiled by Sir Nicholas Stern of the British government and published in 2006, was enormously influential in synthesising these concerns and in convincing politicians, opinion-makers and the general public that climate change was a clear and present danger and that speedy and decisive policy action could still avert the most catastrophic outcomes.[9] The Stern Review argues, in particular, that global average temperatures should not be allowed to rise more than two degrees when compared with temperature levels in pre-industrial times, which means that concentrations of greenhouse gases in the atmosphere should be limited to 450 ppm (parts per million; for comparison, current levels are about 430 ppm which compares to 280 ppm in pre-industrial times). This would require a decrease of 70 per cent in global annual greenhouse gas emissions by the year 2050. The more feasible (but climate-wise more dangerous) objective of 550 ppm with a projected temperature increase of three degrees would require a 25 per cent cut in global greenhouse gas emissions. Much of the political discussion turns around a 500 ppm objective requiring a 50 per cent cut in global annual greenhouse gas emissions by 2050. This would require an effort to arrive at about 80 per cent less greenhouse gas emissions by industrialised countries, including the member states of the European Union.

The implicit factor is that raising the price of carbon emissions raises the price of fossil fuels, in particular of imported oil and gas, and thus discourages demand for these fuels. The massive price rises in electricity markets are partly due to the increase in fuel costs given that gas frequently sets the electricity price during peak hours and coal frequently sets the price during off-peak hours. In other words, in a situation in which energy taxes are almost as unpalatable in Europe as in the United States, action against climate change is an acceptable policy to reduce oil and gas consumption.

That said, the introduction of the EU ETS was not an unmitigated success. Its first phase, 2005–7, was marred by wild swings which severely tested investor confidence. After a high of more than 30 euros per tonne of CO_2 in April 2006, prices essentially fell to zero as it became increasingly clear that the European Commission had allocated more permits to emitters than they needed (see Figure 7.7). Fortunately, forward prices

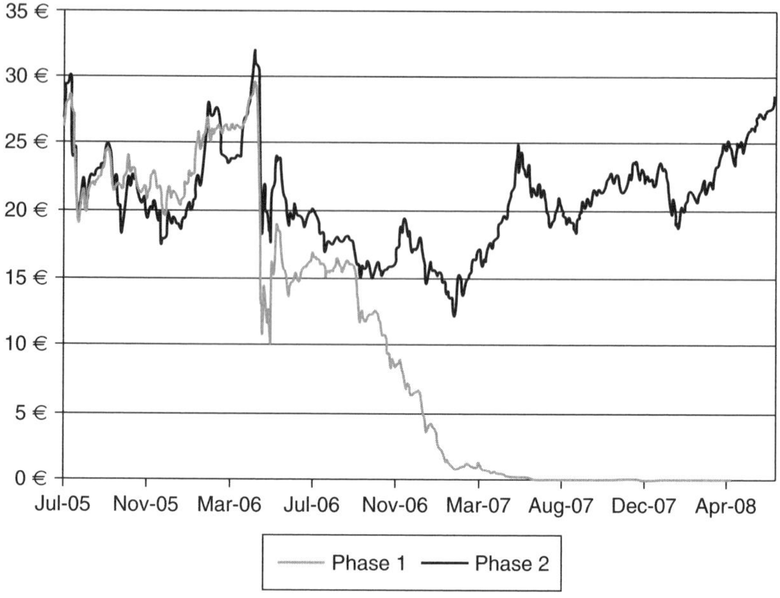

Figure 7.7 Carbon emissions, 2005–2008
Source: Tendances carbone July 2007, Mission climat, Caisse des Dépôts et Consignations.

held up well and the December 2008 contract (pictured in Figure 7.7 for phase 2) quickly decoupled. However, there are great uncertainties also surrounding the second phase, 2008–12. Currently, higher oil, gas and electricity prices all drive the carbon price higher. Such financial correlations are not yet confirmed by solid industrial underpinnings mainly because time frames were too short for investments in abatement technologies to come on board. It is also not excluded that the end of the second period will again witness a period of oversupply. Nevertheless, the EU ETS is the centrepiece of a new carbon economy – to which one needs to add renewable energies, energy efficiency improvements, carbon indices, dedicated funds, investments in carbon capture and storage and many other related efforts – in which Europe plays a leading role.

The European Union is currently mapping out its strategy for the so-called 'third phase', the post-Kyoto period from 2013 to 2020. At the heart of this strategy is the second legislative package on energy and climate published on 23 January 2008.[10] Together with the third legislative package on the liberalisation of gas and electricity markets, it provides a policy-making framework for achieving the EU's famous triple objective to reach a 20 per cent reduction of greenhouse gas emissions, a 20 per cent increase in energy efficiency (both from 1990 levels) and a 20 per cent share of renewable energy in total energy consumption by 2020. While the overlaying of only partly convergent objectives complicates matters from an economic point of view, the rhetorical alliteration of '20s' has captured the public imagination and has developed its own legitimacy.

The second legislative package is composed of seven legislative proposals that neatly capture all the key orientations of Europe's future energy and climate policy. Its immediate political objective is to show that Europe is serious about its announced objectives. Once implemented, it will profoundly influence Europe's energy future.

1. Two texts relate to greenhouse gas emissions reductions by 2020. The first contains a proposal for the politically sensitive issue of burden sharing between the twenty-seven EU member states. The second is a revision of Directive 2003/87 regulating the European carbon market. It contains, in particular, the controversial proposal to auction (rather than to give away at no cost) emissions allowances, a decision that will profoundly affect the profitability of coal-based power producers.
2. One text contains a proposal for a Directive on the promotion of renewable energies. Its most interesting feature is the proposal that member states can comply with their objectives by buying energy produced on the basis of renewables from other member states (in essence establishing an EU-wide market for renewable energy).
3. Another text relates to the objective of improving the EU's energy efficiency with a communication evaluating national energy efficiency plans. While energy efficiency is clearly the 'soft belly' of European energy policy-making, its impact should not be underestimated. It remains linked to a number of structural issues such as transport policy, housing, urban renewal and questions of lifestyle that may, over time, have profound impacts, even if they are difficult to assess in the short run.
4. Two texts relate to the capture and geological storage of CO_2 (carbon capture and storage or CCS). The EU is actively promoting the issue by creating financial incentives as well as the regulatory infrastructures for the one technology able to reconcile coal and climate protection.
5. A final text defines the rules under which public subsidies, such as, for example, specific feed-in tariffs for renewable energies, are compatible with EU provisions against state aid in the common market.

It is rare for a legislative proposal to have such a central position in shaping the realities of the energy market. One needs, however, to understand that implementing Europe's ambitious vision of reconciling climate protection, security of supply and competitiveness will require a sophisticated mix of market and non-market measures. Success will depend on Europe's ability to answer two key questions. First, will the EU be able to resist micro-management and succeed in defining a few clear objectives (such as a 20 per cent emissions reduction), putting simple economic incentives in place (carbon trading, an energy tax) and letting the market get on with it? Second, will Europe be able to engage the outside world in its efforts to reduce emissions and improve security of supply? The two answers boil down to one: success depends on Europe adopting a market-based multilateral approach for *all* energy-related commodities: oil, gas, coal, nuclear, renewables and, of course, carbon.

3 Engaging the world

Once again, energy is becoming a make-or-break issue for Europe. Fortunately, the situation is somewhat less dramatic than in the late 1940s and early 1950s. Nevertheless, in the current protracted identity crisis of the European Union, the European approach to climate protection and its implications for energy is the single most important defining feature of what Europe is all about. The trouble is that others are not listening very well. Currently, European policy-makers prefer to bask in self-righteous solipsism rather than to push for a global extension of their approach. While it is perfectly sound to begin good policy-making at home, the issues at stake ultimately demand global approaches. On multilateral energy market governance or extending and strengthening the Clean Development Mechanism, Europe has been as silent and ineffective as all the other major players. In July 2008, a G8 summit in Japan under the slogan 'Cool Earth' was supposed to address the climate issue but hardly anybody took notice. Europe, as the only group of industrialised countries having committed to significant greenhouse gas reduction, could also forge itself a role as 'honest broker' between developed and developing countries but is held back by its own timidity and lack of imagination rather than by profound internal dissensions. Communicating the European vision to the outside world, however, is crucial if it is ever to become a reality.

For the sake of attaining its climate change objectives and the security of its own energy supplies, Europe must work to improve the multilateral energy trading system and actively promote free, liquid and transparent international energy and carbon markets. There is no harm in bilaterally acknowledging mutual interdependence, such as with Russia or Northern Africa, and in agreeing to work on common projects. However, such bilateral cooperation must not stand in the way of efforts to address the global issues facing energy and carbon markets. It is, however, perfectly legitimate to reflect, in such a multilateral context, about the expediency of border tax adjustments for energy-intensive goods vis-à-vis countries that refuse to be engaged in global efforts to reduce greenhouse gas emissions.

Europe should resist resource nationalism as well as carbon nationalism. However, it should not do so on moral, legal or political grounds. Every effort has to be made to explain that retreating into individual national positions or bilateral agreements constitutes a sub-optimal solution for all involved, particularly in regard to the reduction of greenhouse gas emissions. Europe will need to convince its partners that only a global carbon market will be a viable solution to organise the massive transfers needed over time to convince developing nations to accept even loose emissions targets.

Europe should also quickly abandon its multiple 'neighbourhood policies' that reach as far as the western border of China.[11] 'Projecting soft power' is a notion appealing to former cold warriors searching for new missions, but in practice it distracts Europe from becoming a robust and credible champion of open and global energy and carbon markets.

At the same time, Europe should continue to offer technological, financial and institutional aid freely. Frequently, exporting countries, especially smaller ones, are in dire need of such aid. It is no coincidence that most of the world's energy resources come from politically and economically unstable regions of the world. The role of such help is not to advance 'influence' but to stabilise vital trading partners.

Agreeing on such a radical commitment to an open, market-driven approach would lay the basis for a more successful management of energy and climate issues in the future. It would also contribute to the rationalisation of a debate that remains too often clouded by superficial pronouncements of 'shared responsibilities'. While there are shared responsibilities for securing international markets and defining emissions reduction objectives, defining and defending its position in international markets is a matter for each independent actor: Russian gas exports to Asia are as legitimate as European gas imports from the Middle East. The introduction of moral categories in decision-making on both energy and climate has contributed to a deterioration of relations all round: with the United States on climate issues, with Russia,

Saudi Arabia and OPEC on energy issues. They risk putting new burdens on relations with India and China.

A multilateral trading system with global governance is by far the most likely means to produce benefits for producing and consuming countries alike. Such a global governance structure would need to combine aspects of the International Energy Agency, the UNFCCC, OPEC, the G8 and the WTO. At this point, neither a huge bureaucracy, nor coercive rules are needed. The world, however, needs a permanent forum to discuss, observe and integrate energy and climate issues at the highest level. Europe is the only international actor with the legitimacy to propose such a new structure. Partners in the process of securing and strengthening the international energy and climate trading system must be the United States, China, Russia and Saudi Arabia. In the absence of such a best solution, however, there are a number of existing initiatives and forums, in which Europe can stay involved, continuing to translate its particular vision of the energy and climate issue into global practice:

1. Continued involvement in multilateral organisations such as the International Energy Agency, the World Bank, the UNFCCC and the World Trade Organization (WTO). Wherever possible, energy and climate issues should be integrated into the rule-based dispute settlement mechanisms of the WTO. The Commission should also try to press for a global summit on the multilateral energy trading system.
2. Europe's leadership in the Kyoto process and the creation of the European Emissions Trading Scheme must continue but it must also adapt. Its potentially massive contribution to decarbonising the EU economy further and thus reducing dependence on imported hydrocarbons has not always been sufficiently underlined in recent years. It is an integral part of an EU energy policy and must be part of any foreign policy initiatives in the energy field. Oil-based carbon emissions in the transport sector (both land and air transport) need to be introduced into the ETS as quickly as possible both for reasons of security of supply and for environmental reasons. Only with a coherent policy package – including notably an extension of the EU ETS beyond its borders – will Europe be able to continue to exert the global leadership it has displayed in this increasingly policy-oriented area.
3. Europe also needs to continue to improve the conditions for private investment in supplier countries. Over the next decades, the notion of a 'supplier country' will not only refer to the producers of hydrocarbons but also to the providers of emissions reductions. The European Investment Bank (EIB) and the European Bank for Reconstruction and

Development (EBRD) are supporting energy infrastructures such as the trans-Caspian energy corridor, including the Nabucco gas pipeline, and the project to link the countries of sub-Saharan Africa to the Mediterranean. More important, however, would be creating the legal and technical infrastructures to enable private investment to participate both in energy production *and* greenhouse gas emissions reductions in Third World countries.

4. Europe needs to stay involved in the process of the Energy Charter Treaty. Its focus, however, needs to switch dramatically. First, from an emphasis on 'third-party access' it should switch to an emphasis on defining the rules for foreign direct investment and technology transfer. Second, its scope should be expanded to include energy investments all over the world. The Energy Charter Treaty is a prime example of how bilateral or regional initiatives become bogged down in endless struggles about rent distribution rather than in defining the rules that allow it to be maximised.
5. EU participation in multilateral technical initiatives such as the World Bank's Global Gas Flaring Reduction Partnership (which has both an environmental and a security of supply aspect to it) and the Extractive Industries Transparency Initiative needs to continue. Such initiatives include support for the Financial Action Task Force (FATF) against money laundering as well as broader adoption by EU companies and banking institutions of the Equator Principles promulgated by the International Finance Corporation (IFC) promoting environmentally and socially sound investment. The European Union should also continue and broaden in scope its technical assistance to Russia, East European and Central Asian countries through the TACIS programme that quite naturally leads to joint implementation.

On a more technical and concrete level, there exist a large number of individual initiatives Europe can take to improve the security of its energy supplies and reduce its carbon emissions. The following list is not exhaustive. It is intended to convey an intuitive feeling for the direction of the policy shift advocated rather than insisting on adherence to any specific proposal. Nevertheless, each single proposal constitutes in itself a carefully considered option for a European energy policy. In this spirit, Europe should:

(a) Extend its intellectual leadership in the carbon debate to the energy field and proclaim forcefully its adherence to an open international trading system. Europe should organise for this purpose a large international conference in which the contours of the future interaction of energy and carbon issues are outlined.

(b) Convince consumer and producer countries that price and ability to pay must be the only criteria for access to precious resources and that an open trading system is the best manner to realise the totality of resource rents – and this in both energy and carbon markets

(c) Provide European actors, experts and decision-makers with the legal, technical, informational and economic infrastructure to participate fully in global markets.

(d) Be frank about future high prices for energy. Emphasise that rising prices (i) are a sign of rising global demand which will generate positive spillovers for everybody and (ii) will decrease intensity, augment efficiency and reduce emissions. It should also promote budget-neutral green tax reform.

(e) Assist social groups most vulnerable to high energy and carbon prices but do so in a manner compatible with existing markets and policy orientations. The groups hit hardest by recent price rises – the fishing industry and road haulage – have received subsidised fuels for decades, which has delayed structural adjustment.

(f) Limit financial speculation in both energy and carbon markets by improving market transparency and the energy information infrastructure. While this does not constitute a hedge against permanently higher prices, it can limit speculative bubbles. It should also engage other countries in efforts to improve transparency and disclosure of financial flows arising from energy transactions.

(g) Promote European energy champions capable of competing in world markets. Energy and carbon abatement are risky, capital-intensive activities that require sizeable players on either side of the bargaining table. Future champions will arise in renewables, CCS and emissions reduction projects. It should resist, however, the demand for protection from competition that will inevitably come from those champions and persist with the promotion of free energy markets.

(h) Deepen and broaden the Kyoto Protocol. Europe should continue a forceful policy of reducing CO_2 emissions that, independent of its environmental merits, reduces the demand for carbon-intensive fossil fuels. It should define and adopt the role of an honest broker between developing and developed nations to help shape post-2012 policies.

(i) Create an emergency preparedness mechanism for physical interruptions of gas supplies such as already exists for oil. In the same spirit, each member country should implement minimum requirements for gas storage and participate in a Europe-wide effort on energy solidarity.

(j) Facilitate the construction of two new regasification terminals and an additional refinery to take full advantage of the developing LNG markets and help to remove the global bottleneck in processing heavy and ultra-heavy oils. Both Russia and Northern African countries might be partners in such projects.

(k) Promote a frank and Europe-wide debate on the merits and costs of new nuclear power plants in the European Union. It should organise as speedily as possible a European solution to the disposal of nuclear waste.

(l) Continue to fund research in clean coal technologies, fusion, renewables, carbon storage and nuclear waste disposal. There are large positive spillovers associated with each one of these technologies warranting public involvement.

Europe has the potential to become a key actor in the energy and climate world of the twenty-first century. While different tendencies are currently still at loggerheads, and European policy-makers are not all capable of expressing it with sufficient clarity, the general European policy orientation centred on ambitious greenhouse gas reduction objectives is sound. What needs to be implemented is, first, a more honest acknowledgement that these objectives will imply continued high prices. This acknowledgement should be coupled to a reflection about a budget-neutral 'green tax reform', reducing payroll taxes in exchange for higher taxes on natural resource and energy use. Second, Europe needs to embrace the market once having set out its general policy orientations. This includes global energy and carbon markets which will be the only transmission mechanisms for linking Europe's objectives with those of its global partners.

Notes

1. In this context, the 'Lisbon Strategy', the European Council's commitment in 2002 to making the European Union by 2012 'the most competitive and dynamic knowledge-based economy in the world' has actually been quietly shelved. The eponymous 'Lisbon Treaty' has usefully supplanted it in the lingo of international punditry. A stocktaking, or even less so, a follow-up to the Lisbon Strategy was never announced. The least common denominator of the shelved strategy ('low energy prices') remains an objective, however. On a more general level, it is frightening to see how oblivious European policy-makers seem to be to the fact that these shenanigans – which take for granted that their fixation on the short term and the desire to advance by stealth

are shared by the general public – drive the distrust and lethargy of their citizens. Paradoxically, in the absence of proper European policy-making, the only chance European voters have to vent their frustration is voting against reform treaties designed to advance matters a bit further. It constitutes a classic example of too little and too late.

2. Currently, industrialised countries are in the absurd situation where they tax highly the abundant factor they wish to see employed (labour) and tax moderately the scarce factor whose use they wish to reduce (energy). While it can make some sense to have a light touch regarding energy taxation in the early stages of industrial development due to (a) the positive externalities of access to energy and in particular electricity and (b) the importance of heavy industries, European countries have long left that stage behind. Unfortunately, industrial realities tend to evolve quicker than policy-making mindsets and so-called 'green tax reforms' have not yet progressed from fringe ideas to a central plank of European policy-making.
3. Europe's energy, oil and carbon intensities (the amount of energy, oil or carbon per unit of GDP) have all substantially declined in recent years. This is important. The resilience of Europe's economy in the face of record oil prices (US$147 per barrel at time of writing this book) is due to two factors: (a) the strength of the euro which cushions import prices and (b) the strongly declining oil intensity of Europe's economy. Europe today consumes only 50 per cent of the oil and 60 per cent of the energy per unit of GDP that it consumed thirty years ago. While its economy has roughly doubled, its oil consumption is today slightly below the 15 million barrels per day of 1979. It should also be noted, however, that improvements in energy and oil intensity have decreased in recent years; that is to say that progress is still being made but at ever lower rates. This implies some sort of intrinsic 'speed limit' under 'business as usual' conditions and calls for some scepticism concerning the objective of the European Union of improving its energy efficiency (the inverse of intensity) by 20 per cent by 2020 compared to levels in 1990. Continuing to make progress will require focused policy action in the areas of both fiscal and technology policy.
4. Gas prices (including VAT) in Russia to both residential and non-residential consumers vary between US$35 and US$70 per 1,000 cubic metres depending on the administrative zone. Compared to a world price of around US$235 per 1,000 cubic metres this amounts to a subsidisation rate of between 70 and 85 per cent. Run-away domestic gas consumption is perhaps Russia's (and thus Europe's) biggest energy problem. Recent supply contracts with Turkmenistan (more than 50 bcm per year), however, have somewhat eased the immediate pressure on Russia's export capacity.
5. The latest example is the epic struggle pitting the Commission against vertically integrated producers and a number of member countries on the question of 'un-bundling', i.e. the separation of productive units and transport networks. Next to the creation of an embryonic European regulatory agency, a joint indicator of some planning by network operators, un-bundling is one of the key issues of the third legislative package on gas and electricity markets of 19 September 2007 which is currently being negotiated between the European Parliament, the Commission and member countries. Concluding these negotiations (in addition to those on the second legislative package on

energy and climate of 23 January 2008, see below) is the key objective of the French presidency of the European Union that began on 1 July 2008.

6. The capacity margin is the percentage difference between installed capacity and peak demand. The average number provided above masks great variations among various European countries. In Ireland (21 per cent), Portugal (4 per cent) and the United Kingdom (1 per cent) capacity margins have improved due to massive investments in power generation of 35, 13 and 9 per cent respectively. In Spain instead, the capacity margin decreased by 4 per cent, despite a capacity increase of 5.5 GW, or 8 per cent of total capacity. France, Belgium, Greece and Hungary also remain in fragile equilibrium having depended on imports for more than three months in 2005 (ibid.).
7. Renewable energy has had some very impressive successes, most notably the installation of more than 20 GW of wind-power in Germany. However, considering the cost (more than 3 billion euros per year) and given the facts that other technologies are even more expensive and that large hydropower sites are exhausted, the proposal of the Commission to have 20 per cent of power generation based on renewable energies (up from 15 per cent today) sounds like a very expensive proposition for European taxpayers.
8. Of Europe's 5 billion tonnes of greenhouse gas emissions, 4 billion tonnes are carbon emissions, half of which are included in the EU ETS. Of these 2 billion tonnes, roughly 1 billion is allocated to the power sector, while the other half comes from heavy industries such as steel, aluminium, cement, refining, pulp and paper and glass. Non-point sources such as transport, commerce, services and households are currently not included in the EU ETS but contribute to Europe's emissions reduction objectives through a series of 'domestic projects'. Additional 'credits' (accounting units certifying bona fide greenhouse gas emissions reduction efforts counting towards the reduction objectives contracted under the Kyoto Protocol) can be gained through the Clean Development Mechanism (CDM) and Joint Implementation (JI). These are project-based mechanisms in which countries with obligations under the Kyoto Protocol are credited for carbon-reducing investments in countries without any obligations (CDM) or countries with easily attainable objectives (JI). The former are mainly developing countries, while the latter are composed of the countries of Eastern Europe and the former Soviet Union.
9. See http://www.hm-treasury.gov.uk/independent_reviews/stern_review_economics_climate_change/ for a complete web-based version of the report.
10. The following remarks are largely based on Cruciani et al. (2008).
11. Generalised bilateral initiatives (as opposed to cooperation on concrete projects) are of limited help at best and can be a distraction and a drain on scarce resources. The number of European 'energy dialogues' is currently proliferating without tangible results. Other than the dialogue with Russia, the Commission entertains bilateral initiatives with almost every energy-producing country in the world. SEC(2007)12, a synthetic policy document for high-level decision-makers mentions Memoranda of Understanding with Azerbaijan, Kazakhstan, Turkmenistan and Uzbekistan, a Communication to the Black Sea Council, contacts with OPEC, the Gulf Cooperation Council, Latin America and the Caribbean and a special Africa–Europe Energy partnership. The problem, of course, is *not* that these initiatives exist as part of normal international relations. The problem is that these routine

diplomatic exercises are currently at the heart of the external energy policy of the European Union, an ambition they simply cannot live up to.

References

BP (2008) *BP Statistical Review of World Energy*, June 2008 at www.bp.com

Capgemini (2006) *Observatoire Européen des Marchés de l'Energie*. Eighth edition, October. Paris.

Cruciani, Michel, Keppler, Jan Horst and Kérébel, Cécile (2008) 'Le "paquet énergie et climat" du 23 janvier 2008: un tournant pour l'Europe de l'énergie', *Note de l'IFRI*.

European Commission (2007) DG Competition Report on Energy Sector Inquiry (SEC(2006) 1724), Brussels, 10 January 2007.

Finon, Dominique and Locatelli, Catherine (2006) 'L'interdépendance gazière de la Russie et de l'Union européenne: Quel équilibre entre le marché et la géopolitique?', CIRED, Policy Paper 11, Nogent-sur-Marne.

IEA (2006) *World Energy Outlook 2006*. Paris: IEA/OECD.

IEA/NEA (2005) *Projected Cost of Generating Electricity*. Paris.

Keppler, Jan Horst (2005) *D'un marché binational à un marché européen de l'électricité: Nouvelles régulations des réseaux et ouverture des marchés électriques en France et en Allemagne*. Research Report, Réseau Transport Electricité (RTE), Paris.

Ladoucette, Vera (2006) 'Energy Security in a Fragile World', Special Report CERA Advisory Service, Presentation at Chatham House, 7 November.

Zaleski, C. P. (2006) 'The Future of Nuclear Power in France, the EU and the World for the Next Quarter-Century', *Perspectives on Energy*.

8
Energy Finance: the Case for Derivatives Markets

Delphine Lautier and Yves Simon

For several years, prices and volumes on energy derivatives markets have been increasing at a tremendous rate. Such sustained growths naturally give rise to questions. Should we worry about such a development? Has it gone too far? Are derivatives markets really characterised by a high leverage effect, by opacity and liquidity problems? Do all these markets and transactions really respond to a need? Should we restrain the transactions of speculators in such markets, before they introduce excess volatility, capable of destabilising the underlying physical markets? This chapter proposes answers to these questions, or at least part of an answer, whenever it is possible. It focuses on derivatives markets,[1] and more specifically on energy derivatives markets.

1 Energy derivatives markets: an overview

We first present the historical background of energy derivatives markets. Then we comment on their recent evolution, and compare them with traditional financial assets.

1.1 The creation of derivatives markets in the energy industry

The history of energy derivatives markets is similar to the history of other derivatives instruments. Whatever the underlying asset concerned, the same rule prevails: derivatives markets are always created when the volatility of the physical asset appears, or becomes important.

The first energy derivatives instrument was launched in 1978, on a petroleum product. This was not by chance; indeed, at the end of the 1970s, the second petroleum shock definitively indicated the end of stability of petroleum prices. Moreover, there was no threat of an oligopoly action on a petroleum product, as could have been the case with crude oil. During the next ten years, the fear of volatility led to the creation of derivatives instruments for all of the important petroleum products, at first in the United States, then in Europe: heating oil[2] (1978), gas oil[3] (1981) and crude oil[4] (1983). 1983 is an important milestone as the two futures contracts on crude oil are, in 2008, the two energy contracts that are the most intensively traded worldwide. The 1990s mark the appearance of derivatives markets for natural gas (1990) and electricity (1996). This corresponds to the period succeeding the deregulation of the underlying markets. The movement was once again initiated by the United States and followed a few years later by Europe.

1.2 The recent evolution of energy derivatives markets

The growth rate of energy derivatives markets is tremendous. In 2007, transactions on energy futures markets progressed at a rate of 28.61 per cent. This is the fourth highest rate recorded worldwide in organised markets, whatever the underlying asset considered. With a 42.25 per cent rise recorded in the activity in futures markets, individual equity is in the leading position.

Among energy derivatives, crude oil stands in a special place. As shown by Table 8.1, the futures contracts on crude oil negotiated on the Nymex are in the first place, with 121.53 million contracts exchanged in 2007.

Table 8.1 Top 15 commodity contracts (by number of contracts, in millions)

Rank	*Contract*	*2006*	*2007*	*% change*
1	WTI Crude Oil Futures, Nymex	71.05	121.53	71.04
2	Soy Meal Futures, DCE	31.55	64.72	105.15
3	Brent Crude Oil Futures, ICE Futures	44.35	59.73	34.69
4	Corn Futures, DCE	64.98	59.44	−12.13
5	Corn Futures, CME	47.24	54.52	15.41
6	WTI Crude Oil Futures, ICE Futures	28.67	51.39	79.22
7	No. 1 Soybean Futures, DCE	8.9	47.43	433.13
8	White Sugar Futures, ZCE	29.34	45.47	54.96
9	Rubber Futures, SHFE	26.05	42.19	61.98
10	High Grade Primary Aluminium, LME	36.42	40.23	10.47
11	Strong Gluten Wheat Futures, ZCE	14.68	38.98	165.62
12	Soybean Futures, CME	22.65	31.73	40.09
13	European Natural Gas Options, Nymex	19.52	29.92	53.32
14	Natural Gas Futures, Nymex	23.03	29.79	29.34
15	WTI Crude Oil Options, Nymex	21.02	28.40	35.13

Source: Futures Industry Association.

If we add the 51.39 million contracts negotiated on the Intercontinental Exchange (ICE), the transactions volume becomes 172.92 million! Let us keep in mind that the volume of this futures contract is 1,000 barrels. Moreover other energy derivative instruments occupy the third, sixth, thirteenth, fourteenth and fifteenth places: energy derivatives instruments are thus very important in the commodity futures industry. Last but not least, futures contracts for electricity do not appear in Table 8.1.

Another important feature of energy derivatives markets is their volatility. As Table 8.2 illustrates, commodities at large appear as more volatile than traditional financial markets and, among commodities, natural gas shows the highest price fluctuations.

These different characteristics, combined with a sustained rise in the prices of the underlying physical markets, explain why energy derivatives markets have been at the centre of attention for several years. Recently, these markets have attracted new operators, like hedge funds or institutional investors. This is not a real surprise: most energy markets are now mature; they can be considered as free of liquidity problems, at

Table 8.2 Volatility comparison (annualised volatility, in %)

Market	*2007*	*2006*
Interest rates (money market)		
Eurodollar	17.1	10.3
Euribor	9.5	10.7
Interest rates (government bonds)		
10-year Treasury notes (US)	05.2	03.8
Bunds (Germany)	04.1	03.8
Equity		
S&P500	15.9	09.7
Euro Stoxx 50	15.5	14.4
Foreign currencies		
British pound	06.9	07.6
Euro	06.1	07.2
Commodities		
Crude oil	29.7	26.4
Natural gas	47.2	62.2
Wheat	33.7	29.5
Corn	32.3	28.3
Copper	32.9	38.5
Aluminium	22.1	32.2

Source: Futures Industry Association.

Table 8.3a Volume of exchange-traded futures and options (millions of contracts)

	2006	*2007*	*% change*
Equity	7,330.71	10,308.74	40.62
Interest rate	3,193.41	3,740.87	17.14
Currency	240.05	334.71	39.43
Commodity	1,091.02	1,398.12	28.15
Other	4.36	4.23	–3.06
Total	**11,859.27**	**15,186.67**	**28.03**

Source: Burghardt (2007).

Table 8.3b Notional amount of over-the-counter (OTC) derivatives (US$ billions)

	Dec. 2006	*Dec. 2007*	*% change*
Equity	7,488	8,509	13.63
Interest rate	291,582	393,138	34.83
Currency	40,271	56,238	39.65
Commodity	7,115	9,000	26.49
Credit default swaps	28,650	57,894	102.07
Other	39,740	71,225	79.23
Total	**414,845**	**596,004**	**43.67**

Source: Based upon Bank of International Settlements reports.[5]

least for the shortest delivery dates (see section 4 for more detail). This is precisely the kind of market that speculators appreciate best. Nowadays, these markets are even considered as a new asset class for portfolio management which, as we will see later, might create some difficulties.

To complete this overview, let us not forget that, among derivatives at large, the commodity markets still occupy a modest place, in organised exchanges as well as in over-the-counter markets, as Tables 8.3(a) and 8.3(b) show. Even with a tremendous growth rate, commodity derivatives are far less important, in the derivatives industry, than interest rates. Moreover, as the gap between these different classes of asset is very high, energy derivatives will remain for a long time in their present place.

2 The organisation of energy derivatives markets

A first step towards the understanding of the functioning of energy derivatives markets is to explain what a derivative is. Then, it is important to distinguish organised and OTC markets. Today, there is a progressive fusion of organised and over-the-counter markets. This evolution began in energy derivatives markets. Step by step, it is spreading to all other underlying assets.

2.1 What is a derivatives product?

Derivatives instruments are financial contracts whose price is *derived* from that of an underlying asset such as exchange rate, interest rate or commodity. The nature of the underlying asset may be very variable. Sometimes, the underlying asset is not a physical or financial or even a traded asset. Financial markets have recently witnessed, for example, the appearance of derivatives markets on climate or credit risk. Another interesting point is that the underlying asset can be another derivative! This kind of construction is widespread in energy derivatives markets where the underlying asset of options – a specific category of derivatives instruments – is very often a futures contract, which is another derivative. Such a phenomenon naturally nourishes the fear of a systemic risk in derivatives markets, which can be thought of as a set of Russian dolls.

Once this general presentation has been made, we can ask what are the most commonly used derivatives instruments? The answer is: forward, futures, swaps and options.

A forward contract is a private agreement negotiated between two counterparts. It aims to exchange a given quantity of the underlying asset – let us say, for example, 100,000 barrels of crude oil – at a fixed time in the future. Such a contract defines: the volume of the merchandise, its quality, its delivery place, its delivery date and the forward price of the transaction. It generally leads to a physical delivery and is traded on the OTC market.

Futures contracts are the most important energy derivatives products. They may be defined as standardised forward contracts. The former are negotiated in organised markets, whereas the latter are used in over-the-counter markets. The standardisation concerns each characteristic of the transaction except for its price (which is a futures price). The degree of standardisation is usually extremely high (see Box 8.1 for an illustration of the American crude oil futures contract). This rigidity in the specifications of the futures contract is compensated for by facilitating its transfer between various counterparts and avoiding the problems of liquidity and delivery (these are quite frequent with forward contracts). A consequence of the standardisation is that the transactions on futures contracts rarely lead to a physical delivery. Futures contracts are purely financial instruments.

Box 8.1 The standardisation of the American crude oil futures contract: the Light Sweet Crude Oil contract[6]

- **Trading unit**: 1,000 US barrels (42,000 gallons)
- **Price quotation**: US dollars and cents per barrel
- **Trading hours**: Open outcry trading is conducted from 10:00 a.m. until 2:30 p.m.

Electronic trading is conducted from 6:00 p.m. until 5:15 p.m. via the CME Globex®trading platform, Sunday through Friday. There is a 45-minute break each day between 5:15 p.m. (current trade date) and 6:00 p.m. (next trade date).

- **Trading months**: Crude oil futures are listed nine years forward; consecutive months are listed for the current year and the next five years; in addition, the June and December contract months are listed beyond the sixth year.
- **Delivery**: F.O.B. seller's facility, Cushing, Oklahoma, at any pipeline or storage facility with pipeline access to TEPPCO, Cushing storage, or Equilon Pipeline Co., by in-tank transfer, in-line transfer, book-out, or inter-facility transfer (pumpover).
- **Deliverable grades:** Specific domestic crudes with 0.42 per cent sulphur by weight or less, not less than 37° API gravity nor more than 42° API gravity.

The following domestic crude streams are deliverable: West Texas Intermediate, Low Sweet Mix, New Mexican Sweet, North Texas Sweet, Oklahoma Sweet, South Texas Sweet. Specific foreign crudes of not less than 34° API nor more than 42° API.

The following foreign streams are deliverable: UK Brent and Forties, for which the seller shall receive a 30 cent per barrel discount below the final settlement price; Norwegian Oseberg Blend is delivered at a 55 cent per barrel discount; Nigerian Bonny Light, Qua Iboe, and Colombian Cusiana are delivered at 15 cent premiums.

Swaps are the most important derivative products: they reassemble at least 50 per cent of the total transactions on OTC derivatives worldwide, whatever the underlying asset considered. Swaps are also private agreements negotiated between two counterparts. They lead to the exchange of floating and fixed prices, at regular intervals. There is no physical delivery with a swap. Most of the time, these instruments are used for a long-term horizon. For example, in the gas market, there have been swaps with a commitment as long as twenty years. Lastly, swaps may be considered, from a financial point of view, as portfolios of forward contracts. The former insure a protection against a recurrent price risk, whereas the latter are particularly useful when hedging a punctual risk.

Options, as opposed to forward, futures and swaps contracts, give the right and not the obligation to buy (call option) or sell (put option) the underlying asset. This right may be used at or before a specified expiration date and at a fixed price, which is usually called the strike. As a result of the flexibility associated with this right, this instrument is more costly than are firm derivatives like futures and forward contracts. Options are also fundamentally asymmetrical assets: whereas the buyer of the option has a right, the seller undertakes an obligation; moreover, the value of the option is non-linear, as it corresponds to a maximum value of zero (when the right is not used) and a positive value. Lastly, options may be traded in organised markets as well as in OTC markets.

2.2 Organised versus over-the-counter (OTC) markets

Derivatives markets can be separated into two different but complementary categories: organised and OTC markets. This distinction must be clearly understood, up until the time when it eventually disappears, because it gives a very useful tool to deal with the new complexity of the organisation of energy derivatives markets.

2.2.1 Organised markets and exchange-traded derivatives

Organised markets are generally characterised by their centralisation around a commercial exchange, where all trade occurs (this characteristic is becoming less important nowadays as electronic trading and quotations grow in importance). They are also characterised by the presence of a clearing house.

The clearing house fulfils two economic functions: the management of credit risk through the mechanism of initial margin and margin calls (see Box 8.2) and the management of the market's liquidity through the centralisation of the transactions. Liquidity is also insured, as mentioned before, by the extreme standardisation of the contracts. Thus, in organised markets, it is easy to find a counterpart. The owner of a futures contract will always be able to sell it easily and quickly – which is not the case for a forward contract.

A last characteristic of futures markets is their transparency: prices are free and publicly available, immediately. This is why an economic function fulfilled by an organised market is, as we will see a bit later, price discovery.

Box 8.2 The management of credit risk through initial margin and margin calls

Initial margin and margin calls are tools used by the clearing house in order to insure the management of credit risk in the commercial exchange.

Margin calls aim to prevent a participant from accumulating a financial loss over time and from becoming unable to fulfil his financial commitments. The clearing house determines every day, at the end of the transactions session, the settlement price that will be used for the valuation of all existing positions in the market. If, on the basis of this settlement price, a participant has lost money since the previous day, he will have to pay to the clearing house the difference between the present and former values of his position: he is 'called at the margin'. This money will go to the participants whose positions' value increased since the previous day (this illustrates the fact that the markets are zero-sum games).

Each day, the transactions on the exchange can only begin once all margin calls are paid. Otherwise, were a participant unable to fulfil his commitment, he would be pushed out of the market. In such a situation, the clearing house uses the initial margin in order to compensate for his losses. Thus, the initial margin is an amount of money[7] which aims to cover the maximal loss an operator may encounter during one day on the exchange. Every participant, buyer or seller, has to put up an initial margin. The level of the initial margin, which is determined by the clearing house, represents approximately 1 to 10 per cent of the value of the exchanged contracts.[8]

The management of credit risk by the clearing house imposes administrative and financial costs on all participants: indeed, they must follow, day after day, the value of their positions on the market, and they have to manage, daily, the cash flows associated with the margin calls. As a consequence, some of the operators may prefer, at least for a certain proportion of their activity, the use of the OTC market. Thus, organised and over-the-counter markets are complementary rather than competing forms of organisation.

2.2.2 Over-the-counter markets

In OTC markets, contracts are entered into through private negotiation. These markets may be considered as decentralised networks of participants, as there is no clearing house. In the absence of an institution collecting all orders – and prices – OTC markets are quite opaque (which does not mean that they do not perform well).

Instruments negotiated in OTC markets closely match the needs of hedgers. The absence of standardisation provides the possibility of ensuring a perfect hedge. It also contributes to the opacity of the market, because instruments and prices are not easily comparable. Moreover, as OTC contracts are specific, the liquidity is low in such markets: whenever the buyer of a hedge changes his mind and decides that he no longer needs protection, he must, most of the time, sell the contract back to the financial institution which formerly sold it to him. When the product is complex and specific, the negotiation can turn to the advantage of the financial institution, because competition is not very high for such products.

The absence of a clearing house also signifies that the participants must, themselves, manage the credit risk associated with their transactions. Each operator is indeed exposed to the risk that his counterpart defaults. The longer the maturity of the commitment, the higher is the risk. Thus mutual confidence becomes very important. Quite often, the market participants know each other, and they are far less numerous than in organised markets.[9]

OTC markets are thus different from organised markets. They are not substitutes for one another: the former are particularly useful for industrialists looking for perfect protection against price risks, whereas the latter are especially important for professionals offering protection in OTC markets, and looking for a tool to cover their residual risk.

2.3 The new frontier in derivatives markets

Recently, the 'landscape' of derivatives markets became more complex, especially in the energy field. More precisely, the frontier between organised and OTC markets has become permeable. The reason explaining such an evolution is that operators are more and more concerned about credit risk. In the energy field, the move towards a relative integration of OTC and organised markets has been initiated by the Intercontinental Exchange (ICE). This exchange gained in importance after Enron's bankruptcy when it first proposed to the participants in the energy markets the activity of electronic brokerage.

As OTC markets are decentralised, the brokerage activity is essential: it conveys information. The presence of brokers improves the efficiency of the markets simply by facilitating the matching of demand and supply. For a long time, brokerage was undertaken by phone. Nowadays, however, people are replaced by electronic platforms.

An electronic brokerage system is especially efficient in a market when the negotiated instruments are somehow comparable (as soon as some kind of standardisation appears). In energy markets some OTC derivatives, like forward Dated Brent or swaps, were quite standardised before the appearance of ICE. In such a situation, it becomes interesting for hedgers to compare the different prices offered by financial institutions; an electronic platform significantly enhances the possibility of comparing. It is virtually unlimited in the information it conveys and gives real time information. Thus ICE naturally imposed itself as an actor that must be addressed in energy markets.

While becoming essential in the energy OTC markets, ICE bought the most important organised market on energy instruments in Europe: the International Petroleum Exchange (IPE). Consequently, ICE quickly became a very important participant in the European energy markets, whatever their organisation. Having acquired, through the purchase of the IPE, the skills and knowledge of a clearing house, there was one more step to be taken: to provide the possibility of managing, through the clearing house, the credit risk associated with OTC derivatives.

Today, ICE does indeed make possible the negotiation of 'cleared OTC derivatives'. The management of credit risk relies on initial margins and margin calls, on the basis of a valuation model which is chosen by the clearing house and which depends, naturally, on the specific derivative instrument under consideration.

Such a system provides great flexibility for the participants in the market. They now have four possibilities. The first consists of choosing a pure OTC product. This may be interesting when the risk to be hedged is very complicated, or when the confidentiality of the deal must be preserved.[10] The second possibility is the choice of an OTC product which is proposed on the electronic brokerage system. This may be interesting when the risk to be hedged is rather standard, thus stimulating the competition between several financial institutions. The third possibility is retaining a cleared OTC product. Such a choice is interesting when there is not enough mutual confidence between the two counterparts as far as the credit risk is concerned. It entails, however, financial and administrative costs due to the initial margin and margin calls. Lastly, it is also possible to trade on the basis of pure futures contracts, which are very liquid.

Thus, since the beginning of the twenty-first century, the organisation of energy derivatives markets has changed dramatically. The various segments of the markets are now much more interlinked than a few years ago. Naturally, such an evolution raises the question of systemic risk. We will answer that question a bit later, at the end of this chapter.

3 The economic functions of derivatives markets

Why do derivatives markets exist? While their first function is the management of price risk, they also make possible speculation and arbitrage, price and volatility discovery, and transactional efficiency.

3.1 Risk management

As mentioned previously, derivatives markets are created when there is volatility of the underlying asset. Their main function is to insure protection against price fluctuations of the underlying asset.[11] Hedgers are usually willing to pay for such a service. Their transactions in derivatives markets do not aim to create a profit.

In derivatives markets, the risk is managed in a specific way, by transferring it to various categories of participants. Usually, indeed, hedgers transfer their risk to those willing to assume it. Consequently, risk is not pooled as in insurance. Nor is it diversified, as in portfolio management. Transferring the risk implies reallocating it among operators. It never means that risk disappears. When somebody earns money in the market, somebody else loses: derivatives markets are zero-sum games, and they should not be compared with casinos.

Frequently, risk is not only transferred from a hedger to another participant in the market. Through standardisation and the very high liquidity of the market, when a hedger gets rid of his risk, it is 'sliced' into small parts and is assumed by several persons. This possibility to distribute the risk among several operators explains why the volume of transactions recorded in derivatives markets may be far more important than the quantities dealt with in the physical market. This is neither a problem, nor the sign of poor functioning of the market. On the contrary, the possibility of sharing the risk with numerous counterparts can reduce the price to be paid in order to obtain protection against price fluctuations of the underlying asset. Naturally, when there is a possibility of distributing the risk among several operators, it is not always easy – especially in 'pure' OTC markets – to know where the risk is and who handles it.

3.2 Speculation and arbitrage

As mentioned before, hedging does not aim to earn money. Conversely, profit is the explicit objective of speculation. This should not be a surprise: speculators take on the risks that hedgers wish to avoid. Thus, they ask for remuneration for this service, which is essential. One could think that there is no need for speculators as, in a derivatives market, hedgers looking for protection against a drop in prices are facing hedgers looking for protection against a rise in prices. The trouble is, if they understand correctly the functioning of the physical market, these two categories of hedgers will have the same kind of expectations concerning the future evolution of the prices; and when the entire market thinks that there will be a rise in prices, hedgers looking for protection against a fall vanish. This is the reason why there is a need for speculators.

It is also important that speculators be numerous in a derivatives market, in order to 'slice' the risk into a number of parts. A speculator indeed 'bets' that the market as a whole is wrong, at least over a very short time span. Whenever his bet turns out to be wrong, the speculator must have the possibility to quit the market quickly. Meanwhile, he will leave room for another short-term bet on a small part of the risk, undertaken by another speculator. Thus speculators earn money while allowing the hedger to find protection against price fluctuations. They have a critical function.

Arbitrage also aims to make a profit. However, it does it in a different way. Arbitrage indeed exploits what is usually referred to as an 'abnormal situation', from an economic point of view. Such an abnormal situation may be due to an unexpected event having a strong impact on prices. As arbitrage is not supposed to be risky (another difference from speculation), the possibility to make a risk-less profit attracts a lot of operators and the abnormal situation disappears. Thus, the presence of arbitragers in a derivatives market is also very important because they insure, among other things, that prices in derivatives markets remain linked with the prices of the underlying asset. In other words, arbitragers allow for the convergence of derivatives and physical markets ... and for stronger links between markets (see Box 8.3 for more details).

Naturally, things are not so simple. Hedging, speculation and arbitrage are not as separated concepts as they may appear in the preceding paragraphs. Sometimes hedging can turn into speculation, for example when a hedger renounces his hedge on a futures contract in order to benefit from a favourable evolution in prices. In certain situations, arbitrage can also turn into speculation. And this vagueness makes the study of hedging, speculation and arbitrage very difficult.

Box 8.3 The role of arbitrage in the convergence between the physical and financial markets

In every futures market, even if this is exceptional, it is always possible to deliver the underlying asset at the expiration of the futures contracts. This possibility, associated with arbitrage operations, insures that there is no disconnection between the financial and the physical markets. Let us illustrate this point with an example.

Suppose that, a few days before delivery, the price in the spot market is 140, and that the futures price of the contract reaching its expiration date is 150. In such a situation, an arbitrager may buy the merchandise in the spot market and simultaneously sell it in the futures market. Meanwhile, he announces to the clearing house that he will deliver the physical product at expiration. Doing so, the arbitrager earns 10 (minus the costs associated with the delivery of the merchandise: transport, storage during a few days, etc). Because this operation is profitable and risk-less, there is an incentive for all the participants to do the same thing. Thus, the physical price rises under the pressure of arbitragers buying the merchandise, while, as a result of sales, the futures price diminishes. Arbitrage operations insure the convergence between the physical and financial markets. They stop only when their profit no longer exceeds their costs.

3.3 Price discovery

A third important function of derivatives markets is price discovery. Derivatives markets (especially futures markets) make it possible to obtain information on prices for different maturities. In futures markets, this information is publicly available. Moreover, as contracts are standardised, prices are comparable. Last but not least, they are reliable, as these markets are characterised by an important volume[12] of transactions.

The information conveyed by futures prices is important because it represents the expected value of the future spot price at the delivery date T, based on the current available information (at t, for example). Naturally, information concerning the market will change between the present date and delivery, and these changes are all the more significant when the distance between these two dates is important. Still, futures prices are considered as the best estimators of future spot prices.

Because futures markets fulfil this informational role, they allow for spatial and inter-temporal allocation of resources.

The role of futures markets in spatial allocation may be illustrated by the crude oil market. There are two important futures contracts in the case of crude oil: the Light Sweet Crude Oil and the Brent futures contracts. Prices of the former relate to the American market, whereas the latter is representative of the European market. Usually, the spread between the two futures contracts is about two dollars per barrel. When this spread increases, it becomes interesting to redirect some tankers from Europe to the United States, for example.

The role of futures markets in the inter-temporal allocation of resources comes from the fact that several futures contracts for different delivery dates are simultaneously traded. The relationship between futures prices for different delivery dates is usually referred to as the term structure of futures prices or as the price curve. The curve may be monotonically increasing or decreasing. It may also be sunken or dumped. All these shapes give information on the way to correctly hedge positions that are held in the physical market. Moreover, they also provide guidelines for production, transformation, storage and even, in certain specific cases, for investment decisions, in particular when the maturity of the longer futures contract is remote. For example, when the nearest futures prices are lower than longer prices – the market is then in contango – there is an incentive for the participants in the physical market to buy the merchandise in order to constitute physical stocks. When, conversely, the price curve is decreasing, the market is in backwardation, and the operators are prompted to sell their inventories.

Thus, futures prices motivate financial operations, like hedging, speculation and arbitrage, but also industrial operations. This is one of the reasons why one may think about derivatives markets as a way to stabilise prices in the physical market through optimal production decisions. Thus, prices established in futures markets are very important because they are used by producers, industrialists and consumers. And more importantly, as futures prices provide a reference to trade quality differentials, they are also used by those, throughout the world, who are not directly involved in the exchange, but concerned about the underlying asset.

3.4 Volatility discovery

Provided that there are options traded in the exchange, the informational role of derivatives markets extends itself to volatility. Volatility is the main determinant of options prices. It is so important that options markets are usually referred to as volatility markets.

There are several possible definitions of volatility. The most common is historical volatility, which is computed on the basis of the fluctuation of past prices over a certain period of time. In the presence of actively traded options, it is possible to compute another kind of volatility, usually called implied volatility. This is the standard deviation which equates the market price and the theoretical price of the option derived from a model. Whereas historical volatility solely incorporates past prices, implied volatility discloses operators' expectations of future volatility (conditionally on the available current information). This information is very important as it gives an estimation of the (past and future) risks associated with the positions held in the market.

3.5 Transactional efficiency

The last important function of futures markets is the reduction of transaction costs. Futures markets give direct and free access to competitive trading, insuring reliable prices. Meanwhile, they create benchmarks that are used as a reference to trade OTC transactions and quality differentials. The existence of such benchmarks reduces transaction costs.

4 Should we be worried about the development of energy derivatives markets?

The sustained development of derivatives markets and their complexity leads to fear and raises questions. Among them, it is possible to note the influence of derivatives markets on physical markets, the systemic risk, and the worry that derivatives markets might be intrinsically dangerous.

4.1 The influence of derivatives markets on physical markets

A recurrent question about derivatives markets is whether their presence creates volatility in excess on physical markets.[13] If it were the case, it would be a real problem, because derivatives markets are supposed to be a response to, and not a source of, volatility in the physical market.

One may consider that derivatives markets create some volatility in excess because they allow for very rapid and easy trades. The standardisation of futures contracts, the presence of a clearing house, and the possibility – with electronic platforms – to trade at any hour of the day, indeed makes it possible to immediately react to information affecting the market of the underlying asset. Moreover, derivatives markets allow for noise trading, which are disturbing actions of operators taking advantage of the existence of such a market to speculate more easily, without special concern for the underlying asset. If that point of view is retained, chances are high that there will be an increase of volatility due to the presence of a derivatives market. However, this is not a real problem if the volatility is mostly an intra-day one, reflecting the promotion of informative signals and the redistribution of risk among various categories of operators who are more or less averse to risk.

More important is the long-run volatility (on a monthly, weekly or even daily basis). This phenomenon may be investigated when a new derivatives market – namely, futures contracts – is created on a specific underlying asset. Such an introduction provides an opportunity to compare price volatility before and after the creation of the derivatives market. Empirical studies on this question[14] usually lead to the conclusion that an abnormal volatility is observed just after the introduction. However, this abnormal volatility disappears quickly, and subsequent introductions of other derivatives – like options, for example – seem to have no effect. Moreover, several studies have examined the influence of derivatives trading on the depth and liquidity of the underlying market. Their analysis quite often reveals a strong inverse relationship between the open interest[15] in futures contracts and the spot market's volatility. This finding indicates that the futures market provides depth and liquidity to the physical market.

Another and more important question is whether, in the long run, derivatives markets become more volatile, more speculative than before. This question became more important during the last decade, as a result of the intensified presence of certain categories of investors such as hedge funds. The untimely and heavy interventions of these actors, using commodity markets for diversification purposes, and considering commodities as a new class of asset, could indeed have an impact on prices.

The presence of such operators may indeed be disturbing because their transactions do not always depend on information related to commodities: they could also arise from a sudden change in the stock or bond markets, for example. Thus, if these speculators invest a lot of money in commodity markets, there is a fear that they induce price moves having no relationship with commodity supply, production, inventory and demand. Moreover, as there is a strong correlation between spot and futures prices, a price shock on the futures price of a specific commodity may spread to the physical market. The more pessimistic scenario foresees contagion to other markets, especially to energy markets, as we will see in section 4.2. However, as far as we know, no serious study has reached such a conclusion. On the contrary, Büyüşahin et al. (2008) show that until now, it has not been possible to prove that portfolio diversification towards commodities has a specific impact on their prices.

4.2 Contagion effect and the integration of commodity markets

Another source of questions about commodity markets is the fact that they are more and more integrated, raising the fear of systemic risk. As a result of tightened cross-market linkages, a shock induced by traders or speculators may spread, not only to the physical market, but also to other derivatives markets. This question has been investigated in various ways. The first is the study of the impact of traders on derivatives markets through the so-called 'herding phenomenon'. The second is the study of spatial and temporal integration.

In 1990, Pindyck and Rotenberg proposed to define herding as a situation where traders are alternatively bullish or bearish on all commodities for no plausible economic reasons. Herding is a possible explanation for the co-movement that is observed on the prices of different commodities, namely their persistent tendency to move together. This co-movement can be observed on quite a broad set of markets. It can partially be explained by macroeconomic variables: the expected inflation, the growth in industrial production, the consumer price index, several exchange rates, interest rates, money supply, etc. However, if the co-movement is in excess of anything that could be explained by these common variables, herding could explain this excess. Several studies have been performed on commodity markets in order to confirm this hypothesis. Pindyck and Rotenberg (1990) conclude that there is herding on commodity prices. The results of Ciner and Booth (2001) on Asian agricultural commodity prices support, on the contrary, the common economic fundamentals assumption.

Another way to deal with the subject of derivatives and their impact on commodity prices is to focus on the spatial integration of commodity prices. Jumah and Karbuz (1999) study the spatial integration in the cocoa market, using prices extracted from London and New York futures markets. Interest rates are found to play a key role in establishing long-run relationships among prices. The authors also show that futures prices adjust more quickly to new information, compared to spot prices. Ewing and Harter (2000) study the co-movements of Alaska North Slope and UK Brent crude oil prices. Their results show that these oil markets share a long-run common trend, which suggests that the two markets are 'unified': there is price convergence in the markets.

Kleit (2001) also examines the spatial integration of the crude oil markets. He studies seven types of crude oils. All of them are light crude oils, in order to reduce problems associated with quality differentials. His results show that oil markets are growing more unified. All these studies are important because spatial integration could exaggerate the possible negative influence of derivatives in commodity markets. This question is especially important for energy markets, as they become more and more integrated: many energy derivatives involve cross-market counterparts. Thus, derivatives can spread the disturbance or crisis in one energy market to another.

The debate on the effects of derivatives trading and the contagion effect is also closely related to the issue of temporal integration. Temporal integration has not been widely studied in commodity markets. Relying on the 'preferred habitat' theory, Lautier (2005) investigates whether the American crude oil market is segmented or not. Segmentation is defined as a situation in which different parts of the price curve are disconnected from each other. The study shows that the crude oil futures market is segmented into three parts. The first corresponds to maturities below 28 months, the second is situated between the 29th and 47th months, and the third consists of maturities ranging from the 4th to 7th years. Moreover, chances are high that this segmentation would evolve through time, because since the date it was launched, the crude oil futures market has matured. Indeed, the American futures market has experienced a sustained growth in its transactions volume, pushing away the boundary of actively traded contracts. This phenomenon is reported in most derivatives markets – and it is especially important, since 2000, in commodity markets at large, and more specifically, in energy markets – and one can expect that segmentation will move to longer maturities in the future.

The question of temporal segmentation has crucial implications. Were the markets perfectly integrated, a shock induced on one part of the curve could spread out to other parts of the curve. There would be a kind of temporal systemic risk. However, the existence of partial segmentation also raises some difficulty. Indeed, segmentation has an impact for financial decisions, particularly for all the hedging and valuation operations relying on the relationships between various futures prices. The efficiency of these strategies can be affected by differences in the information content of futures prices.[16] This is also the case for investment decisions, when they are based on the extrapolation from observed price curves to value cash flows for maturities that are not available in the market.[17] All these operations rely on term structure models of commodity prices.

Such a tool aims to reproduce the futures prices observed in the market as accurately as possible and to extend the curve for very long maturities. However, its use requires the estimation of its parameters, and these may depend on the informational content of futures prices.

4.3 Are derivatives markets intrinsically dangerous?

Apart from the fear of systematic risk, a question remains: are derivatives markets intrinsically dangerous? Once again, the answer is not straightforward. There are some reasons to fear complexity, opacity and liquidity as well as the leverage effect. However, such fears are not necessarily justified.

The presence of a leverage effect is one of the most important characteristics of derivatives markets. Leverage means that there is a difference between the price exposure of a position (measured by the value of the derivative contract) which depends on the volatility of the underlying asset, and the investment required in order to enter into the contract. In the case of organised markets, this investment is materialised by the initial margin, which represents 1–10 per cent of the position's value. In the case of over-the-counter markets, this investment is sometimes equal to zero, or equivalent to the implicit value of the confidence in the counterpart in the transaction. In other words, derivatives lower the cost of taking on price exposure. This is perfectly true. Moreover, in the case of organised markets, the cost is even lower because the presence of the clearing house and the use of electronic platforms strongly reduce transaction costs. Last but not least, the cost is also reduced by the very important liquidity in such markets.

Are there some counterweights to the leverage effect? Part of the answer is yes, part is no. The counterweight, in the case of organised markets, lies essentially in margin calls. There is a possibility of suffering huge financial losses in organised markets. However, a participant cannot stay in the market if he is not able to absorb these losses, day after day. Moreover, there are rules in these markets in order to avoid the possibility, for a trader, of taking a very large position (that is, for example, more than 10 per cent of the market) and to suffer, in one day, dramatic losses. In the case of over-the-counter markets, when the clearing house does not interfere in the transactions, the only counterweights are, first, the mutual confidence among the participants in the market and second, the ability to measure correctly the risk undertaken through transactions involving derivative products. Whenever this confidence relies on bad foundations, or this ability is not as good as supposed, there may be difficulties. Part of the crisis linked with the derivative transactions on subprimes credit is due to risk underestimation. With this point of view, the movement of the frontier between organised and OTC markets is good news. Keep in mind that, when a clearing house interferes in a transaction, it assumes credit risk, chooses the valuation model that will be applied to the transaction, and estimates the price risk associated with the transaction, in order to define the amount of the initial margin.

The second potential problem with derivatives is the complexity of the instruments traded. The markets always offer new products – there is a marketing dimension in derivatives, as well as in other goods – and the functioning of these products is not always perfectly understood by all of their potential users, at least at the beginning (problems usually arise in markets that are not mature). The use of derivatives products clearly requires technical competence, and the more the markets evolve, the more these competences rise. Once again, complexity is more important in OTC markets, for a very simple reason. These markets indeed make it possible to find the perfect hedge; that is potentially very specific protection against price fluctuations. These perfect hedges would not be possible in the case of organised markets, because of the necessary standardisation. However, such a benefit afforded by OTC markets signifies, in return, that the negotiated products are sometimes very complicated. Is this complexity always totally explained by a real necessity to offer a protection against a very specific exposure? We leave this question open.

Complexity of the products, absence of a systematic price reporting system – which would be difficult to obtain and not really useful as the negotiated products are not standardised – transactions based on mutual agreements ... all these characteristics of OTC markets favour their opacity. The lack of transparency in these markets is undeniable, at least for some transactions – remember that in order to facilitate the transfer of certain derivatives products, some of them are standardised, and that this part of the activity in OTC markets is increasing very quickly. For financial institutions offering hedging services against price fluctuations, opacity may be a way to restore commercial margins that are, simultaneously, dramatically endangered by electronic trading and by the severe competition in organised markets. Nevertheless, there is a real need for specific hedging products and it is not absurd that financial institutions are paid for assuming the risk transferred by their client – which is not always easy to hedge, especially when the derivatives products are complex. Moreover, some participants in the markets might appreciate opacity as it gives them the possibility to undertake some transactions discretely. The OTC market makes it possible to realise huge deals which, were they openly known, might have promptly destabilised the market.

As far as liquidity is concerned, there are three problems with derivatives. First, the high liquidity of organised markets is an incitement for noise traders to take positions that might create intra-day volatility. A second problem is that liquidity is not regularly organised over time. Third, OTC markets are prone to illiquidity problems. Noise arises from the action of agents who falsely believe that they possess valuable information about what should be the correct price and trade accordingly. Noise traders include investors using technical – as opposed to fundamental – analysis, trend followers, herders, etc. Although these traders are useful as they serve as liquidity providers for informed investors, they may be responsible for excessive short-term volatility due to over-trading, bad timing of their trades, over-reactions to good and bad news, etc. This point has already been commented on in section 4.1.

In most derivatives markets, the transactions volume is not regularly distributed along the prices curve (this is also true for open interest). Indeed, short-term maturities are highly traded, whereas liquidity may be very poor on long-term maturities. For example, as depicted by Figure 8.1, in the European petroleum market, between 2000 and 2007, the one-month Brent futures contract represents 45.93 per cent of the total volume on the first six maturities, the two-month futures contracts

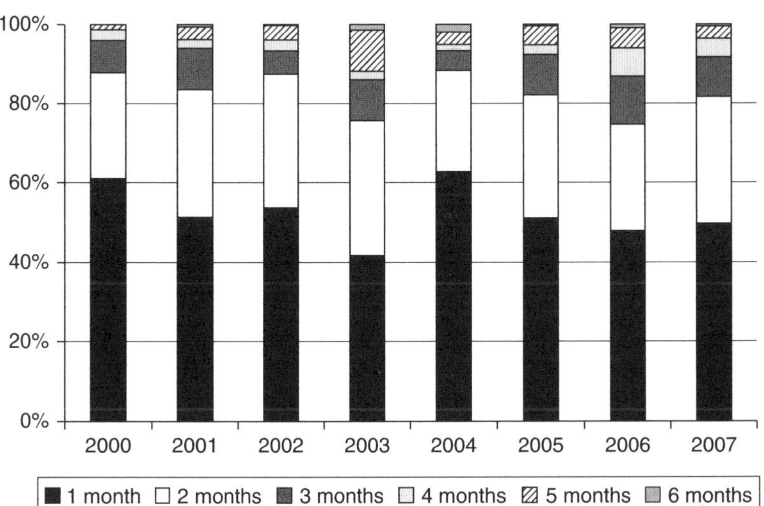

Figure 8.1 Transaction volumes in percentage by maturity on the Brent contract, 2000–2007

amount to 33.91 per cent of the volume, and the three-month contracts, 11.89 per cent. This distribution of transaction volumes might create difficulties for long-term operations in the derivatives markets. Indeed, as long-term prices do not rely on actively traded contracts, one might ask whether these prices are reliable. Three answers can be given to this problem.

First, the prices can be considered as all the more reliable as the integration of the price curve is strong. Were the markets fully integrated in their temporal dimension, the information conveyed by short-term prices would spread rapidly to long-term maturities. Still, it is true that long-term prices are supposed to depend on specific variables which do not affect short-term prices: changes in technologies, inflation, demand pattern, prices for competing energy and changes in environmental constraints are more important for long-term prices than inventories, temporary supply disruption, strikes or seasonality, for example.

A second possible answer is that the transactions volume is not necessarily the best criterion that may be used in order to appreciate the informational content of long-term futures prices. Indeed, when investment banks use long-term futures contracts to hedge their residual risk, one transaction per year may be sufficient – for example to hedge swaps having a one-year periodicity. And investment banks are more and more active with respect to long-term futures contracts, as Haigh et al. (2007) have shown.

Third, it is always possible to cover long-term positions with short-term instruments. Naturally, such a strategy is more risky and requires high technical competence. Just recall that Metallgesellschaft[18] tried to do this in 1994 – and lost US$2.4 billion!

Whereas organised markets are characterised by an irregular distribution of liquidity along the term structure, over-the-counter markets are prone to illiquidity problems, whatever may be the maturity. These problems are directly linked with the organisation of such markets. Don't forget that OTC markets are informal networks, organised around the most important and active operators. At critical times, these operators can vanish or, if they have to stay and propose prices – such is the case for 'market makers' – they might propose prices with no economic sense or, to put it in financial words, with an extraordinarily high liquidity premium. The evolution of the credit market during the summer of 2007 is a perfect example of such a problem.

There are undoubtedly some dangers with derivatives markets, as explained above. However, the most important are probably linked with immature markets, complex products and above all, unadvised traders. Many of the concerns about derivatives markets are due to misunderstanding. Such misunderstanding can create some real difficulties, as was the case, for example, with securitisation in 2007–8. They are sustained by the tremendous volume of transactions, volatility, and by the continual development of such markets. Consequently, there is probably a need to explain, more often and more clearly, what derivatives markets are and how they work. All other problems – namely, the leverage effect, opacity and liquidity – may be overcome, provided that there is no possibility of enforcing the rules governing the markets, especially the organised markets. The vast majority of problems recorded in derivatives markets indeed find roots in fraud or in a lack of control: Metallgesellschaft, Enron, Baring, Société Générale, etc.

5 Conclusion

Since the turn of the twenty-first century, energy derivatives have been growing at a tremendous rate. They are the most actively traded commodities, and also the most volatile. Is it dangerous? By explaining what energy derivatives markets are today, how they work, what their economic function is, and what their possible risks are, this chapter has tried to provide an answer to that difficult question. Naturally, the answer is not straightforward. Moreover, it is difficult to summarise it in a few words.

One important feature of derivatives markets is their complexity. Undoubtedly, derivatives instruments are complex. So is the organisation of derivatives markets, especially in the energy field. This characteristic is harmful, as it may turn away some potential users (for example, developing countries) of energy derivatives markets. Yet, they could be extremely useful, especially when the wealth of these countries relies strongly on energy resources.

The complexity of derivatives markets does not mean, however, that they do not work well. Remember that in 2007 15,187 million futures and options contracts were traded in the world. Moreover, this number represents only the deals in organised markets. Would it be possible to make so many transactions with an inefficient trading scheme? Probably not. Derivatives markets not only work well. They also fulfil economic functions rendering them extremely useful, especially in the energy field, which is characterised by high volatility.

Notes

1. Thus, we will not talk about capital markets, that is, markets aiming at raising funds.
2. The contract was launched on the Nymex (New York Mercantile Exchange).
3. The contract was introduced on the IPE (International Petroleum Exchange).
4. Two contracts were simultaneously introduced on the Nymex and on the IPE.
5. Disclaimer: the BIS does not warrant or guarantee the accuracy, completeness, or fitness for purpose of the BIS material, and shall in no circumstances be liable for any loss, damage, liability or expense suffered by any person in connection with reliance by that person on any such material. The original texts are available free of charge from the BIS website (www.BIS.org).
6. Extracted from the characteristics of the futures contract on Light Sweet Crude Oil, New York Mercantile Exchange.
7. Most of the time, the initial margin is not paid in cash. It corresponds to Treasury Bonds, or to a credit line.
8. Thus, transactions in derivatives markets produce a leverage effect, as there is a need to invest only 1–10 per cent of the value of the position. However, if the level of initial margin which is necessary to manage the credit risk is correctly appreciated by the clearing house, this should not endanger the functioning of the market.
9. In the currency market, which is the second most important OTC market worldwide (after interest rates), the five most important operators represented more than 60 per cent of the transactions in 2007.
10. Even if ICE does not act as a broker or as a clearing house in this part of the market, it is still active: indeed, it offers an electronic system which provides a very rapid and efficient way to confirm the trades of pure OTC products.
11. Thus, derivatives markets are not capital markets. Their first function is not to raise funds. The explanation of the initial margin mechanism (see Box 8.2) also explains that derivatives markets obviously can do a very poor job in recycling petrodollars.
12. In OTC markets, the prices reporting mechanism does not force the operators to disclose their transactions prices.
13. See Mayhew (2000) for an extensive review of the literature.
14. See, for example, Fleming and Ostdiek (1999).
15. The open interest is a very interesting statistic for derivatives markets. It represents the number of contracts still held by operators at the end of a trading day. Thus, it provides an idea of how many participants hold their position more than one day. Arbitragers and speculators usually compensate their position in one day (creating intra-day volatility). Thus, open interest is often considered as a way to appreciate to what extent hedgers are present in the market.
16. For more details on that point see, for example, Keppler et al. (2006).
17. For this kind of analysis see, for example, Brennan and Schwartz (1985) or Schwartz (1997).
18. For more details on the Metallgesellschaft case see, for example, Edwards and Canter (1995).

References

Brennan, M. J. and Schwartz, E. S. (1985) 'Evaluating Natural Resource Investments', *Journal of Business*, 58: 135–57.

Burghardt, G. (2007) 'Volume Surges Again: Global Futures and Options Trading Rises 28 per cent in 2007', *Futures Industry*, May/June issue, Futures Industry Association.

Büyükşahin, B., Haigh, M. S. and Robe, M. A. (2008) 'Commodities and Equities: a Market of One?' Working Paper, January.

Ciner, C. (2001) 'On the Long Run Relationship between Gold and Silver Prices: a Note', *Global Finance Journal*, 12, 2: 299–303.

Ciner, C. and Booth, G. G. (2001) 'Linkages among Agricultural Commodity Futures Prices: Evidence from Tokyo', *Applied Economics Letters*, 8: 311–13.

Edwards, F. R. and Canter, M. S. (1995) 'The Collapse of Metallgesellschaft: Unhedgeable Risks, Poor Hedging Strategy, or Just Bad Luck?' *Journal of Futures Markets*, 15: 221–64.

Ewing, B. T. and Harter, C. L. (2000) 'Co-movements of Alaska North Slope and UK Brent Crude Oil Prices', *Applied Economics Letters*, 7, 8: 553–8.

Fleming, J. and Ostdiek, B. (1999) 'The Impact of Energy Derivatives on the Crude Oil Market', *Energy Economics*, 21, 2: 135–68.

Haigh, M. S., Harris, J. H., Overdahl, J. A. and Robe, M. A. (2007) 'Market Growth, Trader Participation and Pricing in Energy Futures Markets', Working Paper, December.

Jumah, A. and Karbuz. S. (1999) 'Interest Rate Differentials, Market Integration, and the Efficiency of Commodity Futures Markets', *Applied Financial Economics*, 9, 1: 101–9.

Keppler, J. H., Bourbonnais, R. and Girod, J. (eds) (2006) *The Econometrics of Energy Systems*. Basingstoke: Palgrave Macmillan.

Kleit, A. N. (2001) 'Are Regional Oil Markets Growing Closer Together? An Arbitrage Cost Approach', *Energy Journal*, 22, 2: 1–15.

Lautier, D. (2005) 'Segmentation in the Crude Oil Term Structure', *Quarterly Journal of Finance*, 9, 4: 1003–20.

Mayhew, S. (2000) 'The Impact of Derivatives on Cash Markets: What Have We Learned?' Working Paper, University of Georgia, Athens.

Pindyck, R. and Rotenberg, J. (1990) 'The Excess Co-movement of Commodity Prices', *Economic Journal*, 100: 1173–89.

Schwartz, E. S. (1997) 'The Stochastic Behavior of Commodity Prices: Implications for Valuation and Hedging', *Journal of Finance*, 52: 923–73.

Simon, Y. and Lautier, D. (2006) *Marchés dérivés de matières premières*, 3rd edn. Paris: Economica.

9 Winning the Battle?

Jean-Marie Chevalier

In the previous chapters, we have travelled all over the world to see how countries are dealing with the New Energy Crisis. Within the global economy, we have discovered a number of regional dynamics that reflect specific geographical factors: demography, urbanisation, resource endowment, level of income and income inequalities, structure of governance, vulnerability and concern regarding climate change.

For most countries, wealth creation and economic growth are the key strategic priorities but we have, for the time being, a two-speed global economy split between the fast-growing and the slow-growing economies.

Most of the slow-growing economies are OECD countries where growth is often limited by high labour and capital costs, high taxes, costly social protection and structural and behavioural rigidities. Moreover, many countries of this group are dealing with long-lasting financial market turmoil. Among these countries, the United States (5 per cent of the world population, 25 per cent of GHG emissions) might be at a turning point regarding its energy and emissions policy. The twenty-seven countries of the European Union are firmly committed to a process of reducing energy consumption and GHG emissions. However, Europe is facing an 'unsolved triangle' (Chapter 7) between environmental objectives, competitiveness and security of energy supply. These slow-growth economies include roughly 1 billion inhabitants and this population is not expected to grow between now and 2050.

The fast-growing economies are the emerging and developing countries in Asia, Africa and Latin America. They represent a population of 5.5 billion inhabitants that will reach 8 billion in 2050. At that time they will account for 88 per cent of the world population. Most of these countries have low labour and capital costs and low constraints but they are burdened by a number of vulnerabilities. They are vulnerable to climate change both from their own emissions and from the possible effects of global warming. They also face economic and social vulnerabilities often related to poor governance and growing income inequalities. However, they are the engine for the growth of energy demand and they do not feel really concerned by emissions reduction since they consider that developed countries bear the responsibility for the current situation. They are also the engine of the demand for food, raw materials and water. This global demand raises the question of resource scarcity; it raises also the issue of income inequalities between the increasing number of rich who eat more and better and the stagnant and sometimes increased poverty in many areas. Among these countries, the present and potential roles of China and India are crucial.

This picture of the world economy is not very promising when we go back to the equation of Johannesburg (more energy, less emissions). We must keep in mind the IEA scenario described in Chapter 1. The stabilisation case (at 450 ppm) would imply achieving, before 2030, a reduction of CO_2 emissions from the 42 Gt in the reference scenario (in 2030) to 23 Gt (in 2030). In 2009, the prevailing trend is still the reference scenario. The main question of this chapter is to see how these goals are achievable. We also have to keep in mind that the new energy crisis is only part of the global issue. Indeed, the challenge of this century is to provide enough food, water and energy to a growing population without further damaging the planet. Amazingly, the only chapter of this book which mentions food and water is Chapter 4 on energy and economic development.

To provide some answers to these questions, we will examine (1) the dynamics of interdependences which is a new feature of the world economy, (2) the potential role of technology and (3) the need for better and stronger regulation at various levels from local action to global action.

1 The dynamics of interdependences: economics vs. geopolitics

The intensity of interdependences has been considerably reinforced in recent years with the extension of globalisation and the growing importance of emerging countries. It makes the world economy more complex, it facilitates transfers of wealth, it could exacerbate conflicting interests and it makes global regulation more difficult.

1.1 Generalisation of interdependences

The new energy crisis, where energy items such as costs, prices and demand growth are closely related to the issues of climate change, has accelerated the game of interdependences. Here are a few examples:

- Any GHG emission, anywhere in the world, has a local impact but also a global impact since it accelerates global warming and aggravates the risks faced by the most vulnerable countries that are frequently the poorest. Take the case of deforestation which is responsible for more than 18 per cent of global greenhouse gas emissions, more than that attributable to the global transport sector (see Chapter 4). Deforestation, which is still a national problem, has become a global issue. Bangladesh is concerned by deforestation in Brazil and Indonesia.
- The new energy crisis has accelerated the development of biofuels (see Box 9.1). Rising biofuel production in the United States and Europe reduces the demand for oil products but it also reduces the land available for food production and bears some responsibility for the increase of food prices in 2007–8 (IMF 2008). Moreover, the impact of biofuels on the environment is sometimes negative because of the use of fertilisers with the associated N_2O emissions.
- The globalisation of trade and finance has created new interdependences. The third oil shock has promoted crude oil to a 'store of value'. In financial markets, there are permanent arbitrages that concern commodities, stocks and bonds. The picture is still more complex when taking into account the US dollar exchange rate and the level of interest rates.
- The evolution of the world economy is now founded upon the complex game of interdependences. It is also determined by what we called in Chapter 1 the 'dialectic uncertainties of the future': climate change, economics, institutions and geopolitics. The key drivers are: demography with the current geographical and geopolitical shift, the actual effects of climate change and the demand for food, water and energy.

1.2 Economic and financial transfers

Globalisation of trade and finance facilitates transfers of wealth. The game of interdependences might also modify the competitive advantages of nations. The third oil shock and the new energy crisis have brought a new dimension to the functioning of the world economy.

Box 9.1 The case of biofuels

First-generation biofuels include **bioethanol**, which is derived from sugarcane, corn or sugar beets, to be blended with gasoline, and **biodiesel**, which is produced from oil crops such as rapeseed, soybeans or palm, to be blended with diesel oil.

The promotion of biofuels began, after the oil shocks of the 1970s, in order to reduce oil dependence. Trading biofuels soon appeared also as a good tool to support local development through a new demand for agricultural products, thus increasing rural employment and living standards. As climate change became a serious issue, biofuels were seen as a means to reduce the emissions of the transportation sector: when burned, biofuels release only the CO_2 that was taken from the atmosphere when the plant matter was grown.

Until the turn of the twenty-first century, biofuels were not cost competitive compared to petroleum products. Sales relied on public policies which included:

- Incentives for production: subsidies and grants to farmers or to processing units;
- Incentives for consumption: tax reductions on fuels, subsidies or tax reductions for cars;
- Mandatory provisions: minimum blend in fuel, compulsory availability in fuelling stations;
- Standardisation of products: quality of biofuels, 'Flex Fuel' vehicles;
- Public information: raising public awareness;
- Tariff regimes and trade barriers: protection of local producers and nascent industries.

The steady rise in oil prices since 2003 has engendered new enthusiasm for biofuels. In its 2007 *World Energy Outlook*, the International Energy Agency forecasts an annual worldwide growth rate between 7 per cent (reference scenario) and 9 per cent (alternative policy scenario), leading in 2030 to 2.1–3.4 million barrels per day of oil equivalent, displacing 1.8–3.3 per cent of the world oil production.

Such figures have given way to new concerns. Availability of land could be threatened: according to the Food and Agriculture Organisation, the world's arable land used for biofuels could jump from 1 per cent (2005) to 2.5–3.8 per cent by 2030. Biofuels could restrict the availability of water for food production, and food price volatility could become a consequence of the link between food and energy.

The impact of biofuels on the environment is now under review. Except for sugarcane ethanol, the production of the present generation of biofuels requires rather large inputs of fossil fuels. Therefore, these biofuels provide only modest energy and greenhouse gas benefits. Large inputs of fertilisers may lead to increased emissions of the powerful greenhouse gas N_2O. Extension of biofuel land use could destroy lands of high biodiversity, such as rainforests, species-rich grasslands or wetlands. Finally, over-exploitation and inputs to boost yields (fertilisers, pesticides) could damage water and land resources.

A temporary answer to the environment threat is provided by the European Union requirements:

- Biofuels should deliver a high level of greenhouse gas emissions reduction to get public support;
- No biofuel should be made from raw material obtained on land with high biodiversity value;
- International trade should expand good practice, introduce global standards and certification schemes.

In the longer run (after 2020), hopes rely on 'second-generation' biofuels, which are made from lignocellulosic biomass. These biofuels are produced from non-food crops. A larger fraction of the plant is converted into fuel, providing increased benefits with respect to energy and the environment. Yields are expected to be higher and inputs lower, leading to low feedstock costs, while the use of low quality land seems possible.

It seems likely that an economic niche will continue to exist for first-generation biofuels in countries with appropriate climate conditions (Brazil, Tanzania, Mozambique, etc.). The future of second-generation biofuels will depend on technological breakthroughs in automobile engines. In many cases, it may prove more efficient to use the biomass for electricity generation and run 'plug in' hybrid cars rather than producing biofuels.

Source: Michel Cruciani (CGEMP).

1.2.1 The third oil shock as a transfer of wealth from oil consumers to oil producers

Between 2003 and mid-2008, international crude oil prices jumped from 32 to 150 dollars per barrel. This can be considered as a third oil shock

(Chapter 1), very different from the two previous ones because it didn't stop world economic growth. The jump in oil prices resulted in a huge transfer of wealth from consumers to producers. Between 2003 and 2008, OPEC revenues jumped from US$200 billion to US$1,000 billion. Non-OPEC oil-exporting countries such as Russia, Mexico and Canada benefited from similar increases in revenues. It is not easy to track the use of these revenues, especially because the financial sphere is quite opaque. However, we know that part of the money is being dedicated to building, infrastructures, to purchasing weapons and security systems. Part of it is reinvested in the financial sphere. Sovereign funds have become important financial institutions (Chapter 5). They have taken significant capital shares in some financial entities affected by the subprime crisis. Transfer of wealth may result in a transfer of power. A part of oil revenues also feeds corruption and, finally, part of it is reallocated to the population through subsidies for gasoline, natural gas, electricity, health care and education.

On the oil demand side, the effects of the third oil shock are highly differentiated. The United States suffers more than the Eurozone because, in Europe, the high level of taxation and the euro/dollar exchange rate alleviate the shock. For the poorest oil-importing countries, the oil import bill became a real problem for public finance and economic development (Chapter 4). Everywhere in the world, the poor who are using oil products suffer more than the rich. The situation is aggravated by the rising prices of other commodities (food, metals). Higher commodity prices (2007–8) have benefited many emerging and developing economies but they have adversely affected external balances of the net commodity importers. In Africa, vulnerability to commodity markets has disproportionately affected the poor (IMF 2008).

1.2.2 Reshuffling competitive advantages and the balance of power

The whole organisation of the world economy is affected by these transfers of wealth. For the production of goods and services, a number of value chains are now deconstructed and reconstructed on a worldwide basis. Competitive advantages, including labour and capital costs and the price of energy, are key factors of the process. Moreover, it is often more difficult to adapt an old industry than to create a new one. This is illustrated by what is happening in the automotive industry. While the American car industry is in trouble, the emerging Indian and Chinese industries are growing very fast. This illustrates the two-speed global economy. The future of the automotive industry is unclear: What will be the price of oil and biofuels? How strong will be the constraints imposed by climate change? How rapidly will new technologies such as fuel cells emerge economically?

The same questions can be raised for other industries such as airways. What happens to the various global value chains if the price of oil reaches $300? What happens in that case to the tourism industry?

The reshuffling of competitive advantages is mainly an economic question where short-term profitability, economic growth and energy prices are the main drivers. In the long run, however, the most important competitive advantage probably continues to result from education, technology, research and development.

Climate change is directly affected by the reorganisation of the world economy. On the one hand, some developed countries such as the European countries put limits and constraints on their polluting activities. On the other hand, most of the fast-growing economies are looking for growth and wealth. Economic and financial transfers could increase GHG emissions. This evolution is complicated by the fact that emerging and developing countries consider that the developed economies bear the responsibility for the present situation.

These economic arguments show that the resolution of the equation of Johannesburg is more dependent on geopolitics than on economics.

1.3 The geopolitics of conflicting interests

The game of interdependences and transfers makes the world more complex. We are back to the dialectical approach explained in Chapter 1 which emphasises a process of permanent opposition between conflicting interests. In this final chapter the new energy crisis is seen as exacerbating the opposition between a public good – the climate – and a set of private goods generating wealth and money.

1.3.1 The climate as a public good

The international political establishment now recognises the reality of climate change, and accepts the idea that something has to be done. The evolution of the G8 since the 2005 summit at Gleneagles is illustrative. However, the legitimacy of the G8 is now under question. The eight dominant countries represent 65 per cent of the world GDP but only 15 per cent of the world population. They were the participants in the Second World War and no longer represent the present geopolitical situation of the planet. The eight countries are now trying to associate with other participants, such as the Plus Five (China, India, Brazil, South Africa and Mexico), but the two main subjects of disagreement are climate change and the World Trade Organization, two issues that reflect major conflicting interests. There is strong resistance to paying for the management of the planet. The economy has become global but politics is still national. What is true for climate change applies also to the financing of the Millennium Goals.

1.3.2 Tensions over private goods

Private goods generate wealth and money and they are frequently associated with political power. Such an association does not encourage action for climate change. At a global level, demography and the two-speed economy will aggravate scarcity for a number of goods: energy, land, water, food and raw materials. In a context of what we have called 'unleashing capitalism', scarcity will exacerbate economic and political rivalries while growing income inequalities will trigger tensions and social unrest.

1.3.3 Geopolitics of regulation

The new energy crisis and the current functioning of the world economy have a strong potential for generating conflicts and violence. Some of the forthcoming conflicts will be related to the appropriation of natural resources (land, energy, water), some to income inequalities, poverty and famine. Others are directly related to climate change: e.g. forced migrations following droughts or floods. These conflicts could be aggravated by terrorism and the use of nuclear or chemical weapons. Such an analysis, founded upon the dialectics of opposition, calls clearly for the reinforcement of global forms of regulation. The regulation of the world economy appears as a long and difficult political process. We will return to this question in the last section.

2 What can be expected from existing and new technologies?

Technology has a key role in building a sustainable energy future. There is a wide range of new energy technologies and there is also a great potential for using and improving existing technologies in order to improve the efficiency of energy systems. Where technology and technological evolution are concerned, scientists and engineers are fairly confident and optimistic. Politicians, such as George W. Bush, advocate technology – 'the United States is in the lead' – to avoid unpopular constraining objectives of emissions reduction. Economists and financiers are more cautious, mainly concerned by the costs and the financing. The purpose of this book is not to review the possible development of energy technology but to shed some light on the possible contribution of technology to resolving the crisis. Two main ideas are developed: the need for diversity and the priorities to be put forward.

2.1 The need for diversity

The current world energy system has been built on cheap and abundant fossil fuels with no consideration for the effects of energy consumption on climate. Now, climate change matters and technology has a role to play in mitigating its effects. The scientific community does not expect any technological breakthrough between now and 2030. A technological breakthrough would be, for example, a very cheap technology for electricity storage. It could drastically change the organisation of energy systems since intermittent power generation (wind, solar) and also the power generated during low demand periods (at night) could be transferred to higher demand periods. Technological evolution raises two major problems: a problem of cost and the question of the rhythm of evolution.

2.1.1 The difficult question of the costs of technologies and their evolution

From the upstream production of primary energy to the downstream final use, there is a series of costs that are at the same time economic and social (see Chapter 1). The measure and the expected evolution of these costs are subject to a great number of uncertainties. Each technology has its own costs and benefits and there is no technology which can be seen as the ideal answer to the energy crisis. The development of biofuels (see Box 9.1) provides an example where the environmental and economic impacts of current technologies have to be carefully evaluated.

Developed countries are much more sensitive to social costs than emerging and developing countries. This could be another factor in increasing the current economic growth differential between the two groups. A large part of the world population still neglects social costs. Economic growth and cheap technologies are preferred.

The complexity of cost evaluation is a clear invitation for energy diversity. We need all the primary energy sources and we need to experiment carefully with all available energy technologies. Diversity has a value and each country has to promote its own diversity.

2.1.2 *The rhythm of evolution*

The major question is the rhythm at which technology can accelerate the transition towards a low carbon economy. When considering the rigidities, the diversities and the inertia of the world energy systems, one may think that technological answers will be late and slow as compared to the acceleration of global warming. Several elements are crucial for determining the speed of evolution: prices, R&D and energy policies.

Prices. A number of elements, in particular the question of investment, developed in Chapter 1, tend to suggest that the price of oil may remain high. High oil and gas prices coupled with anxiety over supply and increasing concern for the environment should encourage investment in clean technologies which are seen, in California for example, as the greatest opportunity of the century. The price of carbon is another important element, which will be discussed below.

R&D. The Stern Review (2006) underlines the strong decline in public and private R&D spending for the energy sector between 1980 and 2004. The International Energy Agency regularly calls for a major acceleration in R&D for this sector. International cooperation is crucial for developing the technologies of the future. Networks and programmes already exist for nuclear fusion (ITER), hydrogen economy, carbon capture and storage and nuclear generation IV. However, benefits seem to be long-term, as compared, once again, to the evolution of climate change.

Energy policies. Each technology has its own costs and benefits but, among them, the GHG emitting technologies do not pay for the damage to the climate that they are generating every day. One important component of energy policy is to correct this asymmetric situation: to tax or to limit emissions and to set up well-designed incentives for promoting and accelerating low carbon technology options. The pace at which low-emissions technologies emerge depends in great part on national energy policies. Lessons from national experiences and achievements must be available to the world energy community.

The case of nuclear (Box 9.2) illustrates the potential contribution of this technology. However, new developments are slow and nuclear energy will not change drastically the world energy balance. In the case of China for example, the building of twenty-five nuclear plants before 2030 will increase the share of nuclear from 1.2 per cent to 2 per cent.

Box 9.2 The nuclear renaissance: nuclear energy and a low-carbon economy

A few years ago, most energy analysts did not give much credit to nuclear energy development. Even in 2005, the IEA practically ignored nuclear's possible contribution to world energy supply avoiding CO_2 emissions. But by 2007, among the three scenarios presented by the IEA in its World Energy Outlook, the 450 Stabilisation case projected a 12 per cent nuclear share of world primary energy supply in 2030. This would avoid emission of some 6 gigatonnes of CO_2,[a] or some 30 per cent of the 19 Gt[b] of CO_2 which must be avoided if atmospheric greenhouse gas concentrations are to be stabilised at 450 ppm. This 12 per cent corresponds to some 830 gigawatts electric[c] of nuclear power operating in 2030.

Other scenarios, which are bottom-up, analysing the present situation and extrapolating today's trends country by country, result in 740 GWe, 890 GWe or even 1,000 GWe as the possible maximum level of nuclear capacity operating in 2030. These scenarios assume strict emissions controls including a CO_2 price of at least €20 per tonne. Therefore, the range of 740–1,000 GWe nuclear capacity in 2030 seems a reasonable upper limit.

But it is not certain that this high scenario for nuclear energy will be achieved. The main necessary conditions are:

- a very high level of safety (no significant accidents) while maintaining efficient licensing processes;
- regenerating the nuclear workforce (for facility design, construction, operation and for safety authorities);
- investing in new equipment factories and in uranium production and enrichment;
- providing for spent fuel long-term storage and reprocessing as well as repositories for final wastes;
- commercialising fast breeder reactors by the 2040s–2050s to multiply 80-fold the energy extracted from natural uranium (see box, Chapter 1) and minimise final waste by 'burning' long-lived elements;
- developing nuclear energy in a way that does not contribute significantly to the proliferation of nuclear weapons.

If these conditions are achieved, we may consider the more distant future with optimism.

In mid-2008, the IEA published a new study, *Energy Technology Perspectives,* which extends the above-mentioned scenarios up to 2050. This study proposes a group of scenarios called ACT, which would stabilise CO_2 emissions in 2050 at today's levels, and a group of 'Blue' scenarios, which aim to halve CO_2 emissions by 2050. As the study says, 'While ACT scenarios are demanding (difficult and costly), the Blue scenarios require urgent implementation of unprecedented and far-reaching new policy in the energy sector.'

It is surprising that the authors of this report assume addition of nuclear capacity at a fixed level of 32 GWe per year, on grounds that this was the historical maximum reached in the 1970s. In our view, if the world nuclear industry achieves a level of 740–1,000 GWe operating in 2030, it can expand much faster than 32 GWe/y between 2030 and 2050. Moreover, new industry players – China, India, South Korea, and others – which did not exist in the 1970s will become major contributors in 20 or 30 years' time. Some experts predict as much as 59 GWe/y of new construction in the next two decades. Others see the possibility of around 3,000 GWe of nuclear capacity operating in 2050.

Indeed, nuclear energy should be the most competitive way to produce baseload electricity for large grids, provided a minimum value is attached to CO_2 ($50 per tonne suggested in the ACT scenario). The closest competitor will be coal, but even if carbon capture and sequestration (CCS) is successfully developed, coal plants with strict limits on emissions (N_2O, SO_2, particulates, CO_2) will be as expensive to build, if not more so, than nuclear plants and their fuel cycle will be more expensive than the nuclear fuel cycle.

We would therefore suggest than in the ACT as well as in the Blue scenarios of the IEA, the potential role of nuclear energy in mitigating greenhouse gas emissions is strongly underestimated.

In the period to 2030, nuclear energy may make a significant contribution to resolving the dilemma 'more energy – less CO_2', but can by no means resolve it by itself. For the long and very long term – 2050 to 2100 – and assuming that other technology breakthroughs do not make it uneconomic, nuclear energy could make a large, indispensable contribution to world energy supply at acceptable economic and environmental cost.

[a] Assuming that nuclear is replaced by coal without CCS.
[b] The difference between emissions under the 'business as usual' scenario and those under the 450 Stabilisation scenario.
[c] 830 GWe in operation represents more than 600 GWe of new nuclear capacity to put into operation, taking into account the decommissioning of part of the existing capacity of 370 GWe.

C. Pierre Zaleski (CGEMP).
Sources: WNA (2007), Exane BNP (2007), Tourbach (2007).

2.2 Technological priorities

Priorities are not the same for each country but all countries must know what the most promising technological changes are. The IEA regularly tracks energy technologies and their expected evolution. In support of the G8 Plan of Action, the Agency presented global roadmaps, in 2008, of the seventeen technologies that can make the largest contribution, showing what action is needed, and when, to realise their full potential (IEA 2008a) (see Box 9.3).

According to the IEA, these technologies should provide, with an improvement in energy efficiency, the answer to the stabilisation case scenario. The actual contribution of these technologies will depend on costs, prices and the rhythm of evolution. The case of carbon capture and storage (CCS) provides a good illustration. The technology is well known but the cost is still very high and the question of opening sites for storing huge quantities of carbon might pose some problems of environmental opposition. Who is going to pay for the additional cost which roughly doubles the cost of the electricity generated? We are back to the main issue: who is willing to pay for managing the climate?

The roadmaps emphasise what can be done on the supply side and on the demand side. On the demand side, there are huge opportunities for improving energy efficiency by using existing technologies and by developing new technologies. On the supply side, roadmaps are still influenced by the traditional engineering culture of the energy industry which gives preference to centralised systems and economies of scale. In the long run, small-scale decentralised systems might provide other opportunities: the combination of local energy resources (wood, wind, solar, micro-hydro) associated with 'imported' fuels (natural gas, oil), combined production of heat and power, heat pumps, distributed generation. These systems, which combine a number of different technologies, might also open opportunities for bio and nano-technologies.

Box 9.3 Key roadmaps for energy technologies (IEA)

Supply side	**Demand side**
CCS fossil-fuel power generation	Energy efficiency in buildings and appliances
Nuclear power plants	Heat pumps
Onshore and offshore wind	Solar space and water heating
Biomass integrated-gasification-combined cycle and co-combustion	Energy efficiency in transport
	Electric and plug-in vehicles
Photovoltaic systems	H_2 fuel cell vehicles
Concentrating solar power	CCS in industry, H_2 and fuel transformation
Coal IGCC	Industrial motor systems
Coal: ultra-supercritical	
Second-generation biofuels	

(CCS: Carbon Capture and Storage; IGCC: Integrated Gasification Combined Cycle)

3 How to overcome the new energy crisis?

Technology will certainly contribute to overcoming the new energy crisis but, within the current world economic dynamics of *rapports de forces*, the technological answers appear to be expensive and slow, except for the use of existing technologies for improving energy efficiency.

Political action is required at various levels. We have already mentioned the 'repoliticisation' of energy questions (Helm 2007). In fact, the new energy crisis has created a double repoliticisation at two complementary levels: global and national. This is well illustrated by Tony Blair's action after the Earth Summit at Johannesburg in 2002. Tony Blair's government had a key role in accelerating international public awareness of climate change. The academies of sciences of the 8 + 5 countries of the G8 were encouraged to assert the importance of climate change before the G8 meeting at Gleneagles in July 2005. Tony Blair succeeded in persuading George W. Bush that climate change is a serious issue. At the Gleneagles Summit, G8 leaders addressed the challenge of climate change and adopted a Plan of Action. They asked the International Energy Agency to play a major role in delivering the Plan of Action. At the national level, Tony Blair's government set up new incentives for improving energy efficiency and developing renewable sources.

In parallel, the debate on nuclear energy was reopened in order to tackle the issue of reducing future GHG emissions.

Political action is local (the role of local communities or the role of individual states in the United States), national (energy policies), regional (the European Union) and global (Kyoto and post-Kyoto). The range of actions covers all the traditional components of energy policy, but, in terms of climate change, actions have two major components: mitigation and adaptation.

Mitigation is the reduction of GHG emissions. Mitigation has a cost. In theory, the economics of mitigation is founded upon a comparison between an estimate of the expected damages due to climate change and the cost and benefits of emissions abatement (IMF 2008, chapter 4).

Adaptation is a more complex concept. It covers all the various forms of adjustment to climate change. Adaptation may be seen *ex ante*: e.g. the building of infrastructure to protect a city against flood. It can be seen, more dramatically, *ex post*: e.g. migration of people following flood or drought. The question of adaptation has been neglected in the past, because climate change was still a subject of debate (Damian 2007; Oxfam 2007). Now, the financing of adaptation is one of the most difficult challenges of the new energy crisis. The actual cost of adaptation will be high for the most vulnerable people, who are often the poorest. If mitigation is delayed, the cost of adaptation will be higher.

Several factors will determine the vigour of actions. The most important is probably the growing awareness of people, which could be further accelerated by the violence of climatic events. It is public demand and political pressure on governments that will harden action at local, national and global levels. In this matter the international scientific community has a key role. One can already measure people's sensitivity to the issues of climate change. The Yale Center for Environment Law and Policy (2005) has set up an index to measure the ability of various countries to protect the environment over the coming decades. The Environmental Sustainability Index (ESI) uses 76 data sets integrated into 21 indicators which include air quality, biodiversity, water quality, natural resource management, environmental health and energy efficiency; 146 countries are ranked from the most sustainable to the least. The winners are the Scandinavian countries: Finland, Norway, Sweden and Iceland. Other countries such as Japan, Germany, France and the USA are respectively 30th, 31st, 36th and 45th. China is ranked 130th, with particularly low marks for air quality, water quality and the concentration of air pollutants in urban areas. This ranking takes into account the trans-boundary question, when a given country passes on its pollution to others. Such an index enables a permanent benchmarking. It is a tool for putting pressure on the international community with scientific arguments.

3.1 Local, national and regional actions

In the double repoliticisation of energy and environment issues, governments have an important role to play. A dramatic shift in governments' policies is regularly advocated by the IEA. Climate change issues have to be integrated in development strategies, energy policies and corporate strategies. They represent threats and opportunities. The ability of economic policy to help public and private sectors to cope with climate-related risks will be increasingly tested over time. Higher quality institutions and governance strengthen the ability of countries to adapt.

The definition of a national energy policy is firstly determined by the specificity of the country: natural resources endowment, level of income, climate, energy dependence and vulnerability to climate change. Energy policy has four dimensions: the institutional framework in which the roles of the public and private sectors are clearly determined; a supply policy which covers the development of the right energy mix (with taxes and incentives) and also the question of security of supply; a demand policy which is mainly focused on energy efficiency; and a foreign policy for dealing with cooperation, partnership and international negotiations (Chevalier and Méritet 2009). Beyond these general principles, energy policy must now take into account all the interdependences mentioned earlier. An energy policy has to be defined within the general attitude of the country regarding climate change. In Europe, national energy policies are defined within the global European vision of the energy future. Energy policy also has to be closely connected with agricultural policy (biofuels), transport policy, fiscal policy, industrial policy and social policy (issues related to fuel poverty and access to energy). Energy has become the nervous system of the economy.

Without entering into details, we would like to underline three key drivers for local, national and regional action: innovation, energy efficiency and, once again, the need for diversity.

3.1.1 Innovation

We have seen above what can be expected from R&D and technology. The question of technology is most often approached in a segmented way, that is, technology by technology. Innovation is much broader and may consist of combining existing technologies to create new systems: new systems for measuring, controlling and regulating energy flows, new transport systems, new urban systems, and new types of energy procurement and organisation. The new energy crisis is an invitation to innovate.

The new technologies of communication and information (NTCI) have not yet been completely integrated into the current process of deconstruction and reconstruction of energy value chains. Herein lies a great potential for reducing costs, reducing transaction costs, improving procurement and final energy supply, including the associated services. More generally, NTCI provide tools for developing 'energy intelligence' in the design and functioning of energy systems and also for dealing with markets and financial instruments.

3.1.2 Energy efficiency

An improvement in energy efficiency is considered by all energy experts as the first strategic priority to fight climate change. At the G8 Summit, at Hokkaido (Japan), in June 2008, the IEA recommended that G8 leaders adopt and urgently implement a detailed package of measures to significantly enhance energy efficiency (IEA 2008b). The set of recommendations covers twenty-five fields of action across seven priority areas: cross-sectoral activity, buildings, appliances, lighting, transport, industry and power utilities. One may note in this list that buildings account for about 40 per cent of the energy used in most countries. There is a huge potential for energy savings in existing and new buildings.

Implementing these recommendations can lead to enormous cost-effective energy and CO_2 savings. The IEA estimates that, if implemented globally without delay, the proposed action could save around 8.2 Gt of CO_2 per year by 2030. This represents 43 per cent of the emissions reductions targeted for 2030 in the IEA's stabilisation case (see above and Chapter 1). Action on buildings would represent, by itself, more than one-third of the global reduction. In this sector, energy efficiency has to be considered throughout the entire process: planning, design, construction, organisation, management and maintenance.

Studies show that, in most countries, barriers and market imperfection inhibit action on energy efficiency. 'These barriers and failures include hidden and transaction costs such as the cost of the time needed to plan the new investment; lack of information about available options; capital constraints; misaligned incentives; as well as behavioural and organisational factors affecting rationality in decision-making' (Stern Review 2006).

An effective energy efficiency policy implies a very strong and determined government commitment focused on information, education, regulation and financing (IEA 2008b; Stern Review 2006). Government has to be at the same time a leader and a promoter. The actions of the emerging energy services companies seem to be essential.

3.1.3 The need for diversity (once more)

There is no optimal energy system and there is no optimal energy mix. Each country has to find its own way for building its own sustainable energy system by making use of the best practices available. The evolution of automobile use may illustrate the diversity of the future. There is no general model for the future. Some countries will continue to use gasoline or diesel engines. Some will develop ethanol or second-generation biofuel. Others will encourage fuel cell, plug-in electric or hybrid cars.

4 The urgent need for global regulation

The new energy crisis exacerbates the tensions between public goods and private goods, between global welfare and individual interests. The management of the planet and its financing emerges as one of the most important issues of the century. There is clearly a need to reinforce global regulation and create new forms of regulation.

Beyond the equation of Johannesburg and the management of the planet, there are, in the world economy, a number of serious issues that need to be more closely regulated:

- *Money and finances*: the recent financial crisis shows that there is a need to reinforce global regulation and control over financial flows. An increasing share of wealth creation is escaping all forms of control and taxation. The new 'robber barons' of globalisation are using all the facilities offered by tax havens and flags of convenience. Correlated are the issues of dirty money generated by crime, drugs and corruption.
- *Pollution* concerns the GHG emissions and post-Kyoto regulations but also all forms of pollution that result from human activities. Ocean and water pollution are becoming increasingly worrying. All the various components of the nuclear industry, including nuclear proliferation and the trading of nuclear materials, have to be more strictly controlled.
- *International legal frameworks* have to be elaborated and reinforced for a great many issues: sea regulation, property rights, responsibilities of states, companies and individuals, conflicts and dispute settlements. The list is very long and the cost of reinforcing regulation is enormous.

The reinforcement of regulation must be organised through the United Nations institutions. The G8 has rapidly to be transformed into a GX which includes a number of emerging and developing countries. These institutional processes illustrate the political difficulties of reinforcing regulation. This implies long negotiations and high transaction costs.

Before concluding, we would like to mention here the question of the post-Kyoto agenda (Box 9.4) and the associated Clean Development Mechanism (Box 9.5) and then the major issue of carbon pricing.

Box 9.4 The post-Kyoto agenda

The Intergovernmental Panel on Climate Change (IPCC) estimates in its fourth report that the world should reduce its greenhouse gas emissions by at least 50 per cent with respect to 2000 levels by 2050. This translates, for developed countries, to 25–40 per cent below 1990 levels by 2020. Achieving this daunting objective gives the world only a 50 per cent chance of limiting the increase of the average temperature to 2°C by 2100.

In the wake of the IPCC report issued at the end of 2007, governments participating in the negotiations at Bali agreed on a roadmap towards a global agreement to be completed at the Copenhagen conference in December 2009. The Bali Action Plan sets out four issues: ***mitigation, adaptation, technology*** and ***financing***. On these issues, any future international agreement must be looked upon more as a starting point than a conclusion.

As regards ***mitigation***, a key element of the future agreement will be new commitments to reduction targets, as the commitments under the Kyoto Protocol expire at the end of 2012. Unlike the Kyoto provisions, the new agreement should have built-in flexibility. Arrangements may leave room for sectoral agreements; may accommodate stringent targets for developed countries together with a 'crediting mechanism' or non-binding targets for fast developing countries; provide for an improved 'Clean Development Mechanism' mainly aimed at supporting action in the poorest countries; and include adjustment rules and regular reviews. Market tools are now generally seen as the way to lower the cost of emissions reduction. Hence negotiators need to find the right balance between permanent flexibility and the long-term certainty which is required by investors. Specific market tools would be appropriate to prevent deforestation, but effectiveness requires strict rules for monitoring and verification.

The pace of progress will depend on the rate at which developing countries build a robust administrative capacity.

Promoting good governance also remains a major challenge as regards ***adaptation***. Adaptation to climate change is necessary to address impacts resulting from the warming which are already unavoidable due to past emissions. Unlike mitigation, adaptation cannot involve a global market tool such as a carbon price: costs and benefits are different from region to region. Efforts will therefore rely on public funding to gather reliable climate information, develop knowledge on impacts and vulnerabilities, enhance the level of awareness and understanding, improve disaster preparedness and management and upgrade existing infrastructures. International funds will be needed in poor regions, which will suffer the most from the impacts of climate change and have limited adaptive capacity. But rich countries will also have to spend huge amounts of money at home to upgrade their own infrastructures, as was obvious in New Orleans after Hurricanes Katrina and Rita in 2005.

Massive changes in ***technology*** are needed in every country. Technological breakthroughs need R&D investment ('technology push'). Deployment of proven efficient technologies need both economic incentives ('market pull') for developed countries and public support ('technology transfer') for developing countries. Because fast-developing countries cannot be treated the same way as the poorest ones, different tools are being considered. Tools range from a post-2012 Clean Development Mechanism (CDM), which could be strengthened as a vehicle for technology transfer, to a multilateral technology acquisition fund, which could buy out intellectual property rights and make climate-friendly technologies available even for the least developed countries. Within such an international framework, performance-based standards could be harmonised for all traded goods.

When all efforts have been made to reduce to the lowest possible level the overall cost and to select as precisely as possible the destination of the ***financing***, the final question arises: Who shall pay? The fairest answer seems to be that industrialised countries should start paying, thus assuming their responsibilities for the problem. Developing countries will take their share of the burden when they achieve a certain threshold of wealth.

Michel Cruciani (CGEMP).

Box 9.5 The Clean Development Mechanism

The Kyoto Protocol created this project-based mechanism as a tool oriented towards developing countries. This mechanism was considered as an extension within the existing carbon markets, allowing the involvement of developing countries in the process of reduction, but also proposing opportunities for reductions at lower costs.

Indeed, the CDM is a flexibility mechanism giving the right to developed countries to satisfy their reduction commitments through direct actions in developing countries that have ratified the Kyoto Protocol, or by exchanging, in the carbon markets, the obtained Certified Emissions Reductions (CERs).

Thus, based upon the concept of energy efficiency and technology transfer, the CDM represents a real opportunity for developing countries to gain access to sustainable development. But, on the other hand, it allows developed countries (countries from Annexe I) to achieve their targets for emissions reductions by improving the existing industry and energy technology facilities in countries where costs are lower.

Yet, the procedure for the establishment of a CDM project is complex and long-lasting. Every CDM project must follow the path defined by the Marrakech Agreements and monitored by the Conference of the Parties (COP)/ Meetings of the Parties (MOP) and by the Executive Board (for further details see Carr and Rosembuj (2007)).

Several stages allow the creation, the development and the implementation of a project, the verification of its sustainability and also its 'additionality',[a] with the intervention of various experts. Only when the project is settled and its GHG reduction potential is proven, does the Executive Board issue CERs (one CER corresponding to 1 ton equivalent of CO_2 reduction). Of those CERs, a part is transferred to a special account helping developing countries to adapt to climate change and the rest goes to the participants in the project.

Even though the procedure is constraining, CDM flows have increased from US$2.6 billion in 2005 to US$3.3 billion in 2006. CDM investment flows have mostly concerned renewable energy projects and energy efficiency projects.

In 2006, the major participants on the demand side, with a strong commitment to emissions reduction, were the United Kingdom and the rest of the European Union (with 50 per cent and 36 per cent of the market, respectively). The importance of Japan in the CDM market is decreasing as it is willing to implement a more constraining 'cap and trade' system.

The major recipient countries are China and India, representing respectively 61 per cent and 12 per cent of the market in 2006, outstripping Latin America.

However, the present CDM flexibility mechanism needs to be improved and reformed so as to fit better into the complex environmental, economic and institutional situation and to provide an efficient and reliable tool. This necessity results from the confrontation between the European Union's will to fight against global warming more actively (by implementing real actions within its borders and not buying cheaper permits to pollute abroad) and controversial certifying procedures in India, which have recently been reported, that emphasise the risk of corruption.

[a] Its real emissions reduction potential.

Iva Hristova (CGEMP).

4.1 Carbon pricing

There is a large consensus within the international economic community on the necessity to establish a global carbon-pricing framework. The basic idea is that an effective mitigation policy must be based on setting a price for GHG emissions that are driving climate change. According to recent simulations made by the IMF, the overall cost of such carbon-pricing policies – a global carbon tax, a global cap-and-trade system, or hybrid policy – could be moderate and beneficial, provided the policies are well designed (IMF 2008, chapter 4). The IMF concludes that climate change can be addressed without imposing heavy penalties either on the global economy or on individual countries. However, potential adverse consequences such as slower growth, higher inflation and loss of competitiveness must be addressed.

The starting point is probably the EU ETS (see Chapter 7) which has been followed by the development of similar cap-and-trade schemes in both the north-eastern and western states of the United States and in Australia and New Zealand. The linkage between regional trading systems is a first step towards a global carbon market.

The international financial community is attentively following the development of carbon markets. This may provide huge business opportunities. If the United States participates in a global carbon market, it could represent several thousands of billions of US dollars' worth of transactions before 2020. In addition, if emissions allowances are auctioned, it should provide governments with substantial revenues. If carbon pricing is developed, the financial sphere could play an important role of arbitrage and balance between the climate, as a public good, and energy production and consumption, as private goods. A global carbon market would also provide incentives for private sector flows from richer to poorer countries if the CDM is extended and improved.

However, carbon pricing is not enough for resolving the equation (Stern Review 2006; Stern and Tubiana 2008). A global carbon market does not cover all emissions and the institutional establishment of such a market will be a long political process. The issue of deforestation, which accounts for about 17 per cent of global GHG emissions, illustrates how hard it is to take action. The debate on 'Reducing Emissions from Deforestation and Degradation' (REDD) has been revived at the Bali conference (2007). The idea is to provide financial compensation for the reduction of the GHG emissions resulting from deforestation and degradation (Rubio Alvarado and Wertz-Kanounnikoff 2007). Outside of the huge and long-lasting technical problems involved in reaching an agreement, the two major issues are the financing and the capability of local governments to control the permanent activity of deforestation.

5 General conclusion

The new energy crisis is here. Global warming is accelerating and progress in knowledge shows that the impact of climate change could be much more dramatic than initially expected. At the same time the current evolution of the global energy system is on a 'business as usual' trend which is an unsustainable path. Year after year, the situation is getting worse. It is urgent to reduce emissions. On this point, there is a large consensus among international institutions – UN institutions, IPCC, IEA, the World Bank, the International Monetary Fund – and also among the scientific community and many NGOs. However, the same consensus does not exist in the international political community, even if some governments are seriously concerned.

There are two ways to conclude this book: a pessimistic vision and an optimistic one.

Pessimism is founded upon the great difficulty of finding a consensus among nations in order to act effectively. Acting effectively to reduce emissions has a cost. Financing the action is the major stumbling block. When Nicolas Stern asserts that the cost of action for reducing GHG emissions now represents a rather modest investment compared to what would be the cost of inaction for the world economy, he is probably right, but the two categories of cost do not concern the same people. Today very few people and nations are willing to pay for sustainability while the majority of the population does not really care about the costs that will be supported in five, ten or twenty years' time by other generations. Moreover, the demographic shift is giving more weight to the unconcerned people. Kyoto and the post-Kyoto process are first steps but they remain insufficient. There is no agreement for strong collective action and many nations have many reasons for delaying awareness and action. Most political leaders are too busy to give priority to the management of the planet. There is a wide discrepancy between the short-term vision of global capitalism and the long-term public interest. This pessimistic vision results in exacerbation of tensions, rivalries and violence. If there is no collective action, the adaptation to the impacts of climate change will be more costly and more painful. There will be fights for access to energy, water, food and other scarce resources.

The **optimistic** vision is founded upon the human capability to react and to innovate when facing difficult situations. Awareness of the new energy crisis among world citizens is the main factor putting pressure on governments and triggering collective action. International institutions, NGOs and private foundations have an important role to play in accelerating awareness and action. More transparency is required: on information and on physical and financial flows. An example of progress is the Extractive Industries Transparency Initiative (EITI) which requires oil, gas and mining companies to publicly disclose all payments they make to governments and requires of governments, in turn, to report publicly on what they receive from companies. The acceleration of awareness must be accompanied by progress towards better governance and regulation of the world economy. The countries of the G8, but also China and India and other emerging countries, share a responsibility for building new forms of governance and regulation aiming at a sustainable future. Reinforcing world governance means developing a project for the planet: managing the climate, eradicating poverty, increasing access to water and electricity, stopping deforestation and regulating the use of scarce resources. In this project solidarity between nations, people and generations is key.

References

Chevalier, J.-M. and Méritet, S. (2009) 'La politique énergétique' in *Encyclopedia Universalis*.

Damian, M. (2007) 'Il faut réévaluer la place de l'adaptation dans la politique climatique', *Natures Sciences Sociétés* (www.nss-journal.org).

Exane BNP-Paribas (2008).
Helm, D. (2007) *The New Energy Paradigm*. Oxford: Oxford University Press.
International Energy Agency (2008a) *Energy Technology Perspectives: Scenarios & Strategies to 2050.*
International Energy Agency (2008b) *Energy Efficiency: Policy Recommendation.*
International Monetary Fund (2008) *World Economic Outlook.*
Oxfam International (2007) *Adapting to Climate Change: What's Needed in Poor Countries, and Who Should Pay*, briefing paper (www.oxfam.org.uk).
Rubio Alvarado, L. I. and Wertz-Kanounnikoff, S. (2007) 'Why Are We Seeing REDD? An Analysis of the International Debate on Reducing Emissions from Deforestation and Degradation in Developing Countries'. Paris: IDDRI.
Stern, N. and Tubiana, L. (2008) *A Progressive Global Deal on Climate Change*. Yale Centre for Environment Law and Policy.
Tourbach, L. (2007) 'Scénario volontariste de croissance de l'énergie nucléaire: Analyse de l'impact sur la demande en combustible et sur l'émission de gaz à effet de serre' (CGEMP).
WNA Upper Scenario (2008).

Index

Note: Page number in **bold** refer to figures and tables